T0313340

Practical LTE Based Security Forces PMR Networks

To this word, Madam (*the princess Palatine, wife of "Monsieur" the brother of Louis XIV*), which considered to be on the safe side, cries, protests, that except the conduct of her son she ever said nothing or made anything which would displease, and put on complaints and justifications. As she insists most, Madame de Maintenon (*then still the mistress of the king whom she later married*) draws a letter from her pocket and shows it, asking her if she knew the writing. It was a hand written letter to her aunt the duchess of Hanover, to whom she wrote daily, where according to gossips of the court she told her word for word: that no one knew any more what to say about the relation of king and Madame de Maintenon, if it was marriage or an affair; and from there jumped to the business from the outside and those of inside, and extended over the poverty of the kingdom which she told it could not recover.

The postal service had opened it, as it did in most cases and had found it too compromising to just write a summary as usual, and they had sent the full original to the king who had handed it to his mistress.

Saint-Simon, "the court of Louis the XIV", Nelson editor (extracts), page 136, conversation of the 11 june 1701 between the princess Palatine and Mrs.de Maintenon (in this time there was no privacy protection laws and the kings could at will intercept the private correspondances without any protest)

RIVER PUBLISHERS SERIES IN SECURITY AND DIGITAL FORENSICS

Series Editors:

WILLIAM J. BUCHANAN
Edinburgh Napier University, UK

ANAND R. PRASAD
NEC, Japan

Indexing: All books published in this series are submitted to the Web of Science Book Citation Index (BkCI), to CrossRef and to Google Scholar.

The "River Publishers Series in Security and Digital Forensics" is a series of comprehensive academic and professional books which focus on the theory and applications of Cyber Security, including Data Security, Mobile and Network Security, Cryptography and Digital Forensics. Topics in Prevention and Threat Management are also included in the scope of the book series, as are general business Standards in this domain.

Books published in the series include research monographs, edited volumes, handbooks and textbooks. The books provide professionals, researchers, educators, and advanced students in the field with an invaluable insight into the latest research and developments.

Topics covered in the series include, but are by no means restricted to the following:

- Cyber Security
- Digital Forensics
- Cryptography
- Blockchain
- IoT Security
- Network Security
- Mobile Security
- Data and App Security
- Threat Management
- Standardization
- Privacy
- Software Security
- Hardware Security

For a list of other books in this series, visit www.riverpublishers.com

Practical LTE Based Security Forces PMR Networks

Arnaud Henry-Labordère

Professor Ecole Nationale des Ponts et Chaussées, France

LONDON AND NEW YORK

Published 2018 by River Publishers
River Publishers
Alsbjergvej 10, 9260 Gistrup, Denmark
www.riverpublishers.com

Distributed exclusively by Routledge
4 Park Square, Milton Park, Abingdon, Oxon OX14 4RN
605 Third Avenue, New York, NY 10017, USA

*Practical LTE Based Security Forces PMR Networks / by Arnaud Henry-Labordere.

Routledge is an imprint of the Taylor & Francis Group, an informa business

ISBN 978-87-93609-79-2 (print)

While every effort is made to provide dependable information, the publisher, authors, and editors cannot be held responsible for any errors or omissions.

Contents

From the Same Author

Mathematics

[0.1] "Méthodes et Modèles de la Recherche Opérationnelle", Vol. 3 (with Arnold Kaufmann), Dunod 1973, english "Integer and Mixed Programming", Addison Wesley (1976), Russian (MIR, 1975), Spanish (CCSA, 1975), Romanian (1976).

[0.2] "Exercices et Problèmes de Recherche Opérationnelle", Masson, 1976.

[0.3] "Analyse de données", Masson, 1976.

[0.4] "Recherche Opérationnelle", Presses des Ponts et Chaussées, 1981.

[0.5] "Cours de Recherche Opérationnelle", Presses des Ponts et Chaussées, 1995, Vol. 1, Programmation linéaire et non linéaire, théorie des graphes.

[0.6] "Cours de Recherche Opérationnelle", Presses des Ponts et Chaussées, 1995, Vol. 2, Contrôle optimal et optimisation en dimensions infinies, Théorie des jeux.

Télécommunications

[0.7] "SMS and MMS interworking in Mobile Networks", (with Vincent Jonack), Artech Publishing House, 2004.

[0.8] "Virtual Roaming Systems for GSM, GPRS and UMTS", Wiley, 2009.

[0.9] "Virtual Roaming Data Services and Seamless Technology Change", River Publishers, 2014.

[0.10] "Cours d'interception et protection des communications mobiles", FPFL, 2016.

Naval History

[0.11] "Les 4 couleurs du Surcouf", AFHEMA, 2017.

List of Figures

List of Tables

1

Introduction

Quality control procedures, certifications, company ethic charts, belligerent titles such as strategy director in publicity or oil companies, consultants who write tenders by copy-pasting without understanding anything of their compliance questions, standards, all these Anglo-Saxon origin new terms, in a word all this gibberish should be suppressed to leave our talents express themselves, while avoiding the flow of currency which at the end will allow the Americans to buy even the museum paintings of our great statesmen. By having few TV channels, there will be savings and the young will read more literature, culture, history, science. They will draw the inspiration for new ideas useful to their country and to them while developing their taste and respect for the culture and history of other peoples.

André Malraux, Les chênes qu'on abat, 1971, conversations with General de Gaulle.

General de Gaulle correctly forecasted the end of many local computer and telecommunication European industries with the advent of standards. VoLTE and IMS, IoT networks as the rest, are now standardized and all implementations should comply with and face international competitions. Among those who did not start their career in the 1960s, who does remember CII, Télémécanique, Intertechnique, SINTRA, ICL, or Ferranti in UK, Nixdorf or Siemens computers in Germany, Olivetti in Italy, even Bull in France, restricting now to the confidential albeit vital super computers' market?

Let me give an explanation of the title which uses the term "security forces" and the book objectives. It concerns the new LTE-based private mobile radio networks of the security forces, law enforcement agencies (lawful interception), police, firefighting, and associated vital importance operators (VIOs, energy, transport) which will replace proprietary technologies. These networks are designed to provide the security forces with improved large-band data services, push-to-talk and proximity Services (ProSe direct calls without a network) compared with the proprietary Tetra, Tetrapol, and

V25 narrow-band systems (not more than 2G hence unable to transmit large volumes of visio). The objective is also to reduce the public investments by the auxiliary use of public mobile networks (MNOs) with a *resilience possibility* to complement the private coverage provided by "tactical networks," the reserved PMR fixed infrastructures, and also the agreed sharing of the visited VIOs' networks.

This then does not concern "public safety" and emergency call systems which are for ordinary users. There is just a brief description of the VoLTE/IMS case, as this is an opportunity to explain VoLTE/IMS for person-to-person calls which the public security forces may add in the future to their networks.

The main IP group telephony services of the security force mode are quite different from 3G calls or SIP calls. The user clients (their telephone) are permanently connected to an application server (AS; the MCPTT AS also called CGE AS). They press a button to talk to the main group they belong to, and all the members of the group (all those attached to the same MCPTT AS) receive the call. Multicasting may be used to allow push-to-talk or high-quality video sharing (one camera, many receivers) without saturating the radio resources.

This is a very particular area of the 4G technology for which worldwide vendors may not be better than smaller companies from various countries specialized in PMR and able to include all or some of the elements required, more effectively than larger companies. Hence the quote heading this chapter.

Another characteristic is the usage of LTE-based "tactical LTE networks" in a transportable "box" which allows deploying rapidly an autonomous system when the public networks could be saturated. How can they be federated when several public safety groups are simultaneously deployed?

It is clear that 4G will quickly replace 2G and 3G for all "circuit services" voice and SMS with the drive to VoLTE. In France (2017), all operators without any tariff change offer VoLTE, just because the deployment is cheaper and allows a full coverage without the complexity of CSFB. Also new operators have chosen to deploy only 4G. However, a few years are necessary so that all roaming services are opened for all the 4G protocols using Diameter. The book assumes a fair knowledge of LTE networks and concentrates on the lesser known specialized topics concerning PMR networks for public security forces.

Chapter 2 presents new architecture ideas allowing to detail the particularity of the PMR services and consequently proposes an architecture for the "federation" of several tactical networks or the backup use of public networks (with the use of multi-IMSI SIM cards). This subject is operationally important and not covered in other books. It assumes an overall knowledge of 4G standards' mechanisms: roaming, handover, and MBMS (Multimedia Broadcast Multicast Services) which is very detailed in Chapter 6. The reader can jump between the two chapters to understand the details.

Proximity services exist in the previous PMR architectures Tetra or Tetrapol. This is a pure "talkie-walkie" service where the terminal makes direct calls without needing an operation full 4G network. This means that all the terminals should also include a base station function. This is not ready in 2018 as it requires new chipsets. It is currently possible to offer a Prose version which uses only the isolated base stations of a network without a full core network accessible.

The subsequent chapters will detail the specific and lesser known 3GPP features which are used in Chapter 2. At the end, the reader assumed to be familiar with 3G and 4G independently of the public safety case should have acquired the missing knowledges. We start by explaining the features used for the individual isolated tactical network. In each chapter, an introduction links its content to the operational needs of Chapter 2 which is the overall architecture introduction in order to give a very structured book.

Chapter 3: Geo-localization of PMR group members and monitoring of their effective quality of service with the ECID method.

For geo-localization, 3GPP control plane methods or SULP user plane from OMA could be used without difficulty as the security forces have a terminal that can be supplied and configured by their administration. They are standard alternatives to the better known proprietary client-based protocols although finally it is the GPS in the terminal which provides the location measurement. There is no reason that PMR should rely on proprietary geo-localization methods.

I covered extensively control plane location methods in 2G and 3G in previous books [0.7], [0.8], [0.9], and [0.10], which also explain the control plane location methods in 4G using the LPP or LPPa protocols over RRC. There are many interesting optimization mathematics behind it which were addressed by my more than 17 years of research on the LBS topic. Here we present the use of ECID as a most effective tool to read and centralize the "3GPP Measurement Reports" of the terminals which allow having their received signal strength for specific change in the QoS configuration of the tactical network.

The user plane location methods such as secured user plane location (SUPL) are not covered as *there is no implementation difference with 2G or 3G.* All the others use the Diameter protocol for control plane instead of MAP/SS7.

Chapter 4: Choice of the SIM card type for PMR or M2M networks and automatic profile switching possibilities

This is not addressed in other books and is probably the most important choice to provide the possibility of resilience of the PMR network by one or several MNO 4G networks. It could have been the first chapter if the subject had not been so specialized. Should they choose the eUICC type SIMs (several security domains, one for each IMSI) designed for M2M or classical USIM cards with multi-IMSI Applets and a single security domain? SIM card vendors will push the eUICC because it keeps a dependence with their subscription management (SM) system for IMSI (and security domain). MNOs candidating to provide the resilience networks will also push the eUICC so that resilience entails a complete switch to their own security domain and they can use all their existing HLR-HSS setup considering the PMR subscribers as ordinary subscribers. There is also a third choice: eUICC "Consumer" which switch of IMSI has to be controlled by an application in the terminal (device) which prevents using commercial terminals.

Also all details are given on the OTA architecture for reading or updating a SIM card: legacy "push" SMS based on the pure IP OTA where the SIM cards connect to the OTA server and use either BIP/CATTP appeared with 3G USIM cards or recent https introduced with VoLTE able cards. With the references and details given, one could address the development of an OTA IP server. Autonomous cars relying partly on remote servers are the frequent business case to justify the use of eUICC, we discuss the data flows from the various sensors and the eventual implementation by a car manufacturer of its own data only Full MVNO network.

Chapter 5: Group communication provisioning by OTA: SMS/Diameter in 4G

In Chapter 2, we see that the architecture requires that the group leader configures a common identifier for the group communication in the group member terminals. To use 3GPP standards, it should be a particular group APN which must be provisioned when a mission group is created. We assume that the provisioning of GPRS profiles including the APN is known in broad terms. All methods must use at least one "boot" SMS-MT. The possibility and details of SMS purely 4G with Diameter are not well known (there is no such thing as circuit fall back for SMS in tactical networks). Hence, the chapter gives the architecture, details, and dependencies (interfaces S6c and SGd).

Chapter 6: Multicast: MCPTT PMR, MOOC teaching, and TV in local loop networks (RTTH)

The architecture explained in Chapter 2 may use multicast for the downlink sending of the data common to a group. A group can be spread over several tactical networks. We have then provided a chapter to detail the 3GPP multicast with the network need extensions to provide the wanted seamless service. For multicast, we include the BM-SC which interfaces with the standard 3GPP BM2 interface to the AS for mission critical push-to-talk (MCPTT) and the multimedia coordination entity (MCE). Developing a standard MBMS EPC system is a major task as it needs a lab with an eNodeB M2-M1 compliant, a working BM-SC and MCE, and a lot of knowledge of the standards. Multicasting is an efficient TV or MOOC (massive online courses) solution to alleviate the waste of public investments in developing large-band FTTH for the benefit of the big content providers (GAFA) essentially.

Chapter 7: Voice and SMS services based on VoLTE and IMS or VoIP-OTT, access through a non-trusted WLAN (WiFi)

Previously, we mentioned that the main group communication tool is PTT. However, there is the additional need for classical telephone calls and SMS with a regular MSISDN, especially if we have civilian security forces who need to call their office, etc., or receive regular calls or SMS. IMS and VolTE are the most natural solutions to add to the group communication using an MCPTT AS.

Voice services use ISUP, Q703, Q850, or SIP to connect to the PSTN. We cover specific particularities to provide the emergency and short code services which are required for the connection of MNOs.

PMR for policemen and firemen must work from fixed offices which are WiFi equipped. It is of interest that they provide a secured access to the PMR central core for data and voice services (using SIP) and, hence we cover their required architecture with an ePDG and an AAA server. We see that the principle is much like IMS for the authentication.

Chapter 8: Legal (or illegal) interception

Even for security forces, there is an interest to intercept and record all the mobility, voice calls, and data for mission analysis (so-called "media data") and also plain monitoring of the police officers, firemen, and users of the PMR networks. The legal interception architecture is very general but applicable. Based on standards, it is readily available and is explained in this chapter. A rather simple architecture to implement the requirements of law enforcement agencies (LEAs) for core networks is presented. This subject concerns more the future PMR networks: it is enclosed because of the interest and the current lack of knowledge of the standards and technicalities among the PMR engineers.

Chapter 9: Diameter-based M2M (LTE-M and NB-IoT) 3GPP services and LoRa

The PMR networks may include devices (cameras, monitors, and spy microphones), and hence there is a need to explain the standard 3GPP LTE-M narrow-band IoT (NB-IoT) common architecture on the core network side, and low-consumption LoRa architecture quite popular in Europe and backed by a written standard (but with some errors in the latest version [9.11]).

For 3GPP, we explain the MTC-IWF which interfaces the MMEs and SGSNs with the HLR-HSS, all with Diameter interfaces. The idea is to include in the EPS all the standard 3GPP interfaces but not service providers' APIs. To illustrate this definition, the service capability exposure function (SCEF) for IoT will not be provided by our core definition as even if it interfaces with 3GPP S6m/Diameter, it is a special development to offer the service providers' interfaces such as SOAP/ HTTP and various APIs.

For LoRa, I have fully explained it with original detailed call flows. The development on LoRa virtual roaming may be exaggerated in the frame of the book's main subject, but it was an opportunity to explain a very novel subject obviously of more interest to carriers than to security forces which will be M2M users only.

Chapter 10: Advanced policy control and charging (PCC) and provisioning

A few years ago when the first 3G–4G PMR networks appeared, it was considered as very important to be able to allocate the quality of service separately to various members of a PMR network. Higher ranks and operational users would get more. The use of different APNs and PCRF allows a general solution. This is supposed to be known and we focus on a special subject, the implementation of non-neutral Internet, where depending on the service recognized by its IP address, the flow is slowed down. This is more general than PMR and is a solution for those countries desiring to privilege the QoS of national services compared to the restrained bandwidth allowed for GAFA and other world services which do not contribute to the financing of the MNOs' infrastructures while making a successful business of its use.

The convenient provisioning of user data profiles in the equipment such as HLR-HSS and PCRF is a practical subject not only for PMR but also for MNOs. Not really a telecom subject but as it is little known, it was decided to include it.

Appendix 11: Detailed traces for the different chapters

In the Appendix, traces of rarely known procedures used in PMR networks are detailed and commented: over-the-air (OTA) provisioning of SIM cards, multicasting, IoT, and also a practical domain name server (DNS) resolution which is the basis of the data services' roaming implementation both in GSM and in M2M (LoRa).

This is a practical book and the traces are extremely useful for anyone trying to understand in depth and attempting to find on the Internet forums someone who knows the answer.

This book is intended for the LTE safety network developers and for PMR operators, for design and problem analysis; it is to be used as a course material. Hence, the detailed call flows and traces included are the most practically valuable part of the book which is then different from others which are purely informative describing the applications without the explanation of the end-to-end operations. Reverting to the order in my book [0.7], the traces are grouped at the end to make the first reading easier before studying the traces for a course.

A large acronym dictionary is given at the end and a comprehensive list of standards and references is given in each chapter.

Conclusion 12: When can LTE replace proprietary PMR and the 5G real utility

The absence of standard terminals supporting proximity direct talk services will keep the proprietary talkie-walkie services as long as 4G or 5G do not supports them, which means two terminals for the PMR users. The evolution of the core network toward 5G is described by a diagram comparing to 4G with the common parts.

Acknowledgments

Several persons of Halys have contributed with their work or ideas to certain chapters: Gilles Duporche, Sébastien Cruaux, Jean-Philippe Cohet, Waël Manaï, Benoit Mathian. For the SIM cards I have used contributions of François Benoit (ATOS-Bull), from Bertand Vague (Gemalto) and Julia Alves (Idemia). Thanks to Eric Smekens (Gridmax) for the discussion UICC vs eUICC for M2M applications and to Thierry Braconnier for the checking of Chapter 4. The analysis of the various network architectures for the automotive connected applications is based on discussions with Georges-Harald Bernard (MVNO Global). For the MBMS (multicast) I had valuable advices from Jean-Marc Guyot and his colleagues of Enensys, for Chapter 7 (IMS and VoLTE) thanks to Alban Couturier(Cirpack) for useful inputs.

2

LTE PMR Networks: Service, Seamless Federation of Tactical Networks, Backup by the Public Operators' Coverage, and Direct Calls

If you have relations, roaming, support, with the MNOs of some countries, Mali, Zimbabwe, Central Africa, Iran, and others, if you need something quickly and well done with a smile, ask the women.

2.1 PMR tactical network elements

We discuss an autonomous mobile network, a "tactical bubble" having its own eNodeB and network core, 4G exclusively, sized to receive up to few hundred users of public security forces of law and order or armed forces.

The mean of communication is mainly the push-to-talk thanks to an AS called MCPTT AS, a particular case of CGS AS (group call system AS). The group calls are therefore $1 \rightarrow N$ from the one who presses the button "Talk" to call its group.

To limit the used bandwidth, especially in video mode, the use of multicast 3GPP is advantageous. It means therefore that only one common "downlink" channel is existing for voice or video reception.

When isolated, the operation is very conventional and comparable to SIP VoIP telephony. Users identify themselves on the HSS and create one session with the PGW through the MCE. The diagram of an isolated tactical network is represented in Figure 2.1.

Users, who are constituting a task force, communicate exclusively with members of the same group under the radio coverage of the eNodeB linked to the EPC (generally only one eNodeB).

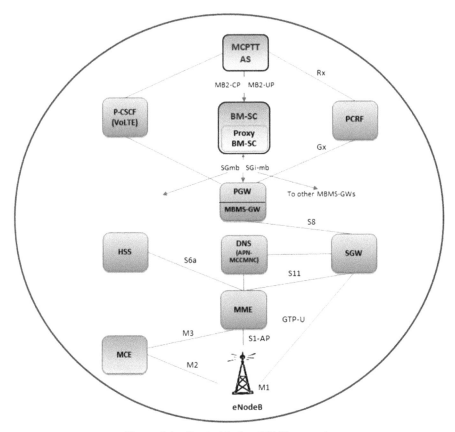

Figure 2.1   Tactical isolated PMR network.

2.2 PMR tactical networks' federation

2.2.1 Operational needs' summary

We want to be able to link several tactical networks (they are the same like in Figure 2.1) between themselves in order that each group could *use the union of their radio coverage* in a transparent way (continuous handover) and *without any manipulation at the level of their termina*l during the mission. Hand-over 3GPP mechanism should therefore be realized in order to switch from a coverage to another one. Group calls, using the multicast, must be available whatever tactical network used. Technical construction of the federation is initiated as soon as a tactical network is connected to one of them – they are the same through an IP link of "backhaul" (cable, satellite link, etc.), arbitrary, which take the central role of a "leader." Therefore, the users of group 1 become visitors of the other federated tactical networks.

With regard to Figure 2.1, the new system in Figure 2.2 perfectly resolves the aim of not interrupting sessions when tactical networks are federated. The system in Figure 2.1 was interrupting sessions to immediately recreate them and was not taking into account multicast. On the other hand, it was holding a localization central HSS for operating to allow incoming phone calls or OTA push from a central server if needed.

Afterward, we are going to add this possibility using the HSS possibility to receive a registration information with the Cx/Diameter protocol (intended to be used for IMS) when users are connecting to their MCPTT Home.

2.2.2 Radio planning and IP addressing of the various federated tactical networks

The different tactical networks use the same radio code, which is recorded as HPLMN and equivHPLMN in the SIM cards. The SIM cards are UICC or eUICC with several "auxiliary IMSIs" eventually, to have a radio coverage resilience in all the available MNOs. In the frequent case when there is no national roaming, a foreign IMSI which has roaming with all the MNOs of the country is a common method (cf. ASTRID in Belgium). The key subject of UICC vs eUICC is covered in Chapter 4.

For radio planning:

- Multicast-broadcast single-frequency network (MBSFN) used for the eMBMS (nothing to do for the eMBMS in case of handover)
- one scrambling code (PCI-*Physical Cell Id* – in LTE, 504 values) different for each eNodeB by tactical networks.

For the IP planning:

Tactical networks can be linked between them; therefore, various tactical bubbles' equipment have different addresses administrated like for a local network. Their various MCPTTs have then distinct addresses allowing terminals to tie to them according to their mission configuration (see Section 2.2.4).

Technically, all tactical networks are similar but one of them, playing the key role, uses an "IP switch" which allows us to make communicating local IP of different virtual equipment of the tactical networks.

- SGWs and PGWs in the preparation and the management of roaming data sessions

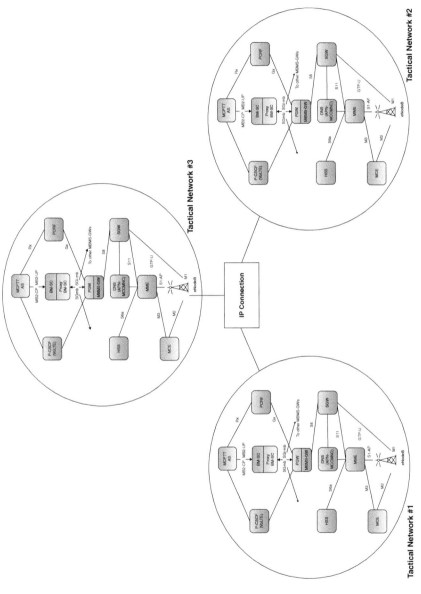

Figure 2.2   Tactical networks interconnected through IP with a tactical network used in central GW IP.

- "Connection monitors" within the tactical networks and BM-SCs for the notification of new multicast areas (the Notify_of_new_PGW) which are therefore going to be created by the BM-SCs.

Various tactical bubbles are connected by an IP cable or by the satellite connection box to the IP switch of a tactical network designated as "central" (which can also have a satellite connection box). If a basic IP switch is used, there is only one local network linking each tactical network (with a proper "IP switch"), without "Gateway IP." It follows that:

- IP communications between local equipment do not come out.
- It is possible for a tactical network to communicate with the SGWs and PGWs of all other ones.

Their "connection monitor" module is going to regularly broadcast the local IP of their PGW in order to broaden multicast coverage areas of different groups.

2.2.3 Radio planning for mobility between tactical bubbles of a federation: Requirements and solution

Users have unmarked IMSIs *not allowing to know from the IMSI to which group they are belonging to*. These groups are formed when a mission is beginning. On the other hand, they have specific APNs of their own group. For the authentication, visited MME is going to use a domain name

$$3gpp.mnc000.mcc208.3gppnetwork.org$$

which does not allow to distinguish the group. The authentication request will then always be addressed to the HSS of the visited tactical network (this is not the classical roaming where it is addressed to the HSS of the HPLMN).

> - They have to be able to use any tactical bubble when isolated or federated, as well as the coverage of any linked tactical bubble for a "Federation." For classical VPLMN roaming architecture –> the HLR-HSS of the VPLMN, diagram for the authentication from HLR-HSS Home is not suitable as it is based exclusively on IMSI. All PMR users are therefore created within the HLR-HSS of all tactical networks. The one of the visited networks is used for the authentication function by S6a/Diameter protocol.

As for the rest, this is roaming and classical handover architecture.

For the "data" service, the APN system, special to each tactical network, will allow any user to have its data session created with the PGW of its home network and keep the same IP, including in case of handover on another tactical bubble.

APNgroupe245.mnc000.mcc208.3gppnetwork.org

- During a mission with its mobile tactical network, at the creation point of a group, every member is logically attached to this network which becomes its HPLMN.
- The phone service is exclusively the "push-to-talk." There are no incoming individual calls in the sense of 3G voice services using the HLR which manages a determined subscriber.

For reception of group calls, subscribers are attached to a group, itself anchored to the PGW and the MCPTT of its tactical network.

2.2.4 Initial configuration of a user to associate with its assigned group

Two methods:

a) At the level of the MCPTT application inside terminals: manual, "the domain of our tactical network is police345" says the group leader, "It is our APN";

b) Alternative VoLTE future at the level of SIM card within used fields for the VoLTE/IMS, by OTA SIM only which highlights before departure the parameter Efp-cscf including P-CSCF address which will be the one of the local MCPTTs. The OTA uses the "build" file with the group members.

To MCPTT clients, it is recommended that they are standardized, regarding the use of the configured APN in the phone (not in hard-code in the MCPTT application), and containing the domain name of the MCPTT above. (It can be automatically in a) the tactical network during the "build" of the group (an USB key with participants' data) compares with the previous and initiates the updating through OTA only the new ones. When connected, they receive the updating and can therefore connect themselves to MCPTT local police345.

- *OTA GPRS* server in the "bubble" (and OTA SIM) if b was used
- MME of the bubble able to receive SMS-MT for the updating

Having specific APNs for each bubble will allow visitor home and handover standard operating although IMSIs are unmarked and do not allow to know the group of a particular subscriber.

2.3 Federation method for $N-1$ concurrent networks with one taking the central role

2.3.1 Architecture description

The principle is unmarked IMSIs but a specific APN for each tactical network and its group.

The APN (specific)–IMSI (unmarked) couple allows any DNS to address the PGW of each tactical network.

The $N-1$ bubbles establish a backhaul IP with the central bubble using the IP connection box of Figure 2.1. Each group has its own MBMS service channels; hence, a maximum of six groups and tactical networks can be federated. The connection can be with:

- A simple IP switch but then all the IPs are in a single network, and their fixed assignment must be centrally planned for all the possible tactical networks which can be federated.
- An IP router, which connects different local IP networks, but this will require its configuration in the field when the tactical networks need to federate.

Groups will be able to visit any coverage and communicate between them. The MCPTT application is composed of two groups:

- my group (own group)
- other groups

All data flows, including "visitors" ones go through the SGW of the visited bubble (instead of PGW Home) which cannot be known as IMSIs, are part of a single pool which *does not allow the DNS to perform an IMSI-based only resolution.* Users can authenticate and register with *any tactical bubble HLR-HSS* as they are all unmarked *with ALL subscribers, including those of other groups. This is a key characteristic of the architecture [2.1] as the IMSIs do not characterize a group, only the APN.*

Customers connect therefore to the "data" through the visited SGW (classical roaming data). And from there connect to their PGW and then to the local address of their MCPTT Home which has the configuration of their own group.

Clients connect then to the "data" through the visited SGW (classical roaming data) and from there connect to their PGW and then to the local address of their MCPTT Home which has their own group configuration.

Following these steps, classical data service in a point-to-point mode (unicast, Figure 2.3) is available.

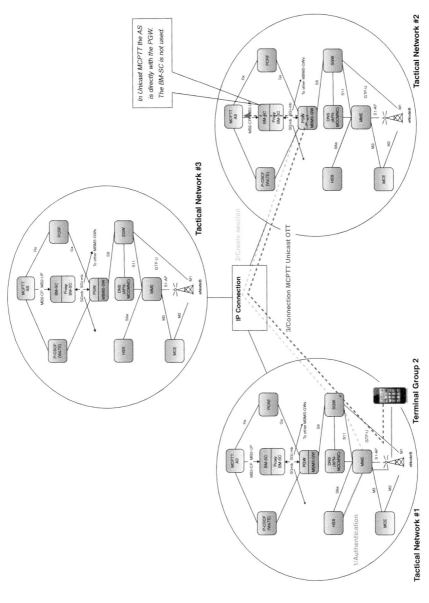

Figure 2.3   Interconnected tactical networks through IP: creation of classical data service (unicast).

Their MCPTT client connects to the MCPTT home server through the backhaul as the APN is configured on each client (only one). Whatever is the visited tactical network, terminals can receive data from applications of their home network. We are now going to add the multicast service.

2.4 Using the multicast for MCPTT and federating MBFSN areas

2.4.1 Introduction to eMBMS

The MBSFN is a communication channel defined in 4G called Long-Term Evolution (LTE) which did not exist in 3G. The transmission mode is intended as a further improvement of the efficiency of the enhanced multimedia broadcast multicast service (eMBMS) service [2.6], [6.1], which can deliver services such as mobile TV using the LTE infrastructure, and is expected to compete with dedicated mobile/handheld TV broadcast systems such as DVB-H and DVB-SH. This enables network operators mobile or fixed to offer mobile TV without the need for additional expensive licensed spectrum and without requiring new infrastructure and end-user devices, as the MBMS is beginning to be supported (2018) by several device vendors which include the middleware necessary for certain TV Codecs (Dash, Flute).

The eMBMS service can offer many more TV programs in a specific radio frequency spectrum compared to traditional terrestrial TV broadcasting, since it is based on the principles of interactive multicast, where TV content only is transmitted in where there currently are viewers. The eMBMS service also provides better system spectral efficiency than video-on-demand over traditional cellular unicasting services, since in eMBMS, each TV program is transmitted only once in each cell, even if there are several viewers of that program in the same cell. The MBSFN transmission mode further improves the spectral efficiency, since it is based on the principles of dynamic single-frequency networks (DSFNs). This implies that it dynamically forms single-frequency networks (SFNs), i.e., groups of adjacent base stations that send the same signal simultaneously on the same frequency sub-carriers, when there are mobile TV viewers of the same TV program content in the adjacent cells. The LTE OFDMA downlink modulation and multiple access scheme eliminates self-interference caused by the SFNs. Efficient TV transmission using similar combinations of Interactive multicast (IP multicast) and DSFN has also been suggested for the DVB-T2 and DVB-H systems.

MBMS for mobile TV was a failure in 3G systems, and was offered by very few mobile operators, partly because of its limited peak bit rates and capacity, not allowing standard TV video quality, something that LTE with eMBMS does not suffer from.

LTE's eMBMS provides transport features for sending the same content information to all the users in a cell (broadcast) or to a given set of users (subscribers) in a cell (multicast) using a subset of the available radio resources with the remaining available to support transmissions toward a particular user (so-called unicast services). It must not be confused with IP-level broadcast or multicast, which offers no sharing of resources on the radio access level. In eMBMS, it is possible to either use a single eNode-B or multiple eNode-Bs for transmission to multiple UEs. MBSFN is the definition for the latter.

MBSFN is a transmission mode which exploits LTE's OFDM radio interface to send multicast or broadcast data as a multicell transmission over a synchronized SFN. The transmissions from the multiple cells are sufficiently tightly synchronized for each to arrive at the UE within the OFDM cyclic prefix (CP) so as to avoid inter-symbol interference (ISI). In effect, this makes due to the absence of inter-cell interference. MBSFN transmission appears to a UE as a transmission *from a single large cell,* dramatically increasing the signal-to-interference ratio (SIR).

To avoid the need to read MBMS-related system information and potentially (SC-)MCCH on neighbor frequencies, the UE is made aware of which frequency is providing which MBMS services via MBSFN or SC-PTM through the combination of the following MBMS assistance information:

- User service description (USD): In the USD (see 3GPP TS 26.346), the application/service layer provides for each service the *TMGI × flow identifier* (which is the identification of a common DL service, a TV program, the listening channel of an MCPTT group), the session start and end time, the frequencies, and the MBMS service area identities (MBMS SAIs, belonging to the MBMS service area (which is the geographical coverage of the service as a list of service area identifiers). In the start of an MBMS session procedure, the mapping of these two-octets SAIs [6.11] to a subframe of a multicast channel is configured in the MBMS capable eNodeBs (see Chapter 6).
- System information: MBMS and non-MBMS cells indicate in System-InformationBlockType15 the MBMS SAIs of the current frequency and of each neighbor frequency.

The MBMS SAIs of the neighboring cell may be provided by X2 signaling (i.e., X2 setup and eNB configuration update procedures) or/and OAM.

In the core network, the common data are transmitted using either broadcast IP or multicast IP (only to a list of eNodeBs), which in both cases avoids the IP traffic duplication.

2.4.1.1 Broadcast mode

The common DL fate is received by all UEs in an MBMS service area. Their application may use them or not. An MBMS bearer is identified by one TMGI. The flow identifier allows different subservices within this bearer and an MBMS session is started by giving the *TMGI x flow identifier* in the GCS action request sent by the CGS AS (MCPPT AS) of Figure 2.1 as well as the list of SAIs of the MBMS service that are for this service. This is the simplest to implement.

2.4.1.2 Multicast mode

The UE must register to the service and obtain keys from the HLR-HSS as described in [2.] using its Zh/Diameter interface.

2.4.2 Attachment of a tactical network in an existing federation: GCS AS-centric architecture

We want to make available the multicast service for a group whatever the visited tactical network by one of its members with a "GCS AS-centric" architecture. Each one of the tactical networks broadcasts ongoing (within few seconds) a new message *Notify_of_new_ PGW added to the protocol SGmb/Diameter* between eMBMS GW and BM-SC. An alternative as GC1 is already proprietary is to use a GC1 message UE → GCS AS to inform that a new cell is visited which is not included in the MBSFN area# 1.

In the current protocol standards MB2 for the GCS AS (MCPTT) or for the BM-SC, there is nothing to inform of a dynamic change of the MBFSN areas; all the configurations are static in the eNodeBs, MMEs, MBMS-GW, BM-SC, and AS; this is why a new "notify" message is required.

This message is send to a multicast IP address[1] that all the BM-SCs listen continuously and which is permanently configured in all the tactical networks to avoid to have any configuration in the field. On start-up, while

---

[1]The broadcast IP address is computed by the operation **OR** between *the bit complement of the subnet mask* and the host IP private address of the BM-SCs. If this is 172.16.0.0/12 and the subnet mask is 255.240.0.0, then the broadcast IPv4 address is 172.16.0.0 OR 0.15.255.255 = 172.31.255.255. The BM-SCs of 1 and 2 listen to this IP and receive the notify new network a proprietary extension of SGmb/Diameter.

being isolated, each BM-SC such as #3 has already created a bearer multicast for its group from its eMBMS GW.

In Figure 2.2, we start from the situation where network 2 is linked yet to the core network 1. The tactical network 3 connects to the core network. Its notification module begins to broadcast the IP of its PGW that the tactical networks 1 and 2 receive. On its side, its BM-SC#3 receives messages of 1 and 2 and proceeds to establish the MBMS bearer of group 3 within the tactical networks 1 and 2 which the coverage is now operable. At the same time, BM-SC#1 and #2 add coverage of #3 to the one of their groups 1 and 2. Each group has now an MBMS service area using all the coverages of all tactical networks.

2.5 MBMS extension of the radio coverage of the new joining tactical network

Consider group#1 to which one wants to add the coverage SAI#2 provided by the coverage of tactical network#2 [2.2]. The GCS AS (MCPTT) of group#1 is informed that SAI#1 is available (proprietary protocol message) as an extension of its MBFSN area which will consist now of SAI#1 and SAI#2 *which in general intersect*, especially for PMR. Chapter 6 explains the details of the protocol used in the creation of a multicast session. An MBFSN area consists of a group of cells within an MBFSN synchronization area (comprises several MBFSN areas) which are coordinated to achieve an MFSN transmission. All cells within the MBFSN area contribute to the MBFSN transmission and advertise its availability with their BCCH channels.

2.5.1 Crude basic federation (cross-copying) active service to another service area

The BM-SCs are unaware that a federation occurred. There is no synchronization of the MCEs and there are potential issues at the boundary of the SAIs which may receive unsynchronized multicast control channels (MCCHs) for group#1 from SAI#1 eNodeB(s) and SAI#2 eNodeB(s). For each MBFSN area, there is a single MCCH channel from each eNodeB.

GCS AS#1 creates a new session with the BM-SC#1 by a GAR request (A#1, SAI#1). It receives the IP and Port of BM-SC# 1 in order to forward it the user plane data of TMGI#1. The multicasted downlink data for group#1 are duplicated by GCS AS# 1 to BM-SC#1 and BM-SC#2 and will be sent then by *two unsynchronized MCCHs*. In the intersection zone, the signal will be degraded as the two MCCHs' content for group#1 should be absolutely identical. This crude approach is then just as a step for the correct architecture, there must be one MBFSN area for each group.

2.5.2 Federated MCEs or central MCE?

One must create one unique MBFSN system for group#1 as well as for the other groups. The management of the *allocation of resources must be coordinated and hence centralized* as in SAI12 = SAI1 + SAI2, the resources may not be enough for the two groups. Some eNodeB vendors say that they include the MCE, but it means only that they interface with M3 with an MME. A MCE *must be centralized* to arbitrate the allocation of MBMS resources to different groups competing for MBMS channels; arbitration means that the allocation when the groups are not federated may be modified.

When federation occurs, the MCEs exchange their contexts so that each is aware of the allocation they already performed for group#1 and #2. An alternative is to address all these contexts to a central MCE as in Figure 2.4 but a distributed design is architecturally more balanced.

GCS AS#1 creates SAI12 = (SAI1 + SAI2) the union of the two service areas. Only GCS AS#1 controls the GC1 of group#1 for all the eNodeBs. There is one MBSFN area: no boundary issue and all the broadcast is synchronized.

2.5.2.1 MBMS LTE channels

There is a one-to-one correspondence between an MBCCH and an MBSFN area. MCH is the new channel introduced in LTE TS 36.300 [6.8]. It concerns the simultaneous transmission of a synchronized *downlink only* signal by several eNodeBs. For the UEs, the signals are superposed and hence increase the signal level.

MCH multiplexes MCCH and one multicast traffic channel (MTCH) which contains the data broadcasted in the eNodeB coverage. The MCCH consists of a single radio resource control (RRC) message which lists all the MBMS services with ongoing sessions. All cells within an MBFSN area transmit an MCCH (except the "reserved cells" excluded from the MBFSN area). MCCH is transmitted at each repetition period and may include changes which are monitored by the UE.

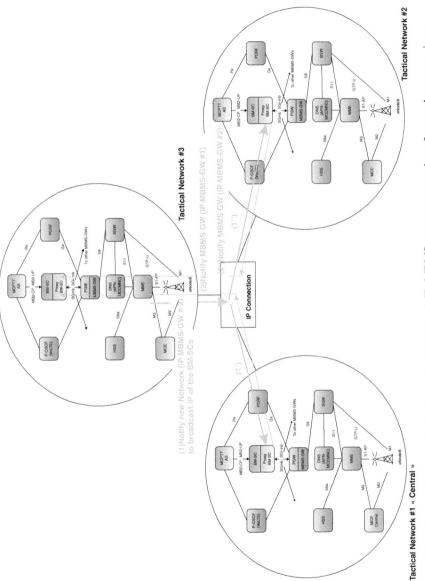

Figure 2.4   Interconnected tactical networks through IP: MBMS coverage extension of groups' areas services.

2.5.2.2 Meaning of "MBMS synchronization," role of the central or coordinated MCE

All the eNodeBs are clock synchronized including GPS; synchronization for MBMS means that the MCCH and MTCH occupy the same subframes sent by different eNodeBs of the same MBSFN area. This also means that the "MCCH-related BCCH configuration" must be the same for all eNodeBs of an MBFSN in the various M2AP messages which contain this information.

BCCH is the usual downlink channel for broadcasting incoming calls, SMS, etc., and MBMS service information since its addition in 4G. All the UEs listen to the BCCH of the radio station they are camped on. BCCH indicates: the scheduling of the MCCH, the modification period, the repetition period, the radio frame offset, and the subframe allocation. These parameters *"MCCH-related BCCH configuration" of the BCCH* are set by the MCE with the M2AP protocol in three M2AP messages (Table 2.1).

The MCCH and MTCH subframes are configured by the MBMS SCHEDULING INFORMATION M2AP Message which contains the physical multicast channel (PMCH) configuration list. The MCE coordinates the transmission from multiple cells in an MBFSN area, each signal being randomly slightly delayed (does not overlap exactly). To correct this, a time guard is added at the beginning and end of the MTCH subframes to prevent an ISI so that an ISI-free signal may be recovered with a fast Fourier transform. The MBMS-enabled eNodeB (M1 interface) must implement the SYNC protocol so that MTCH subframes are sent at the same time so that the random delay is limited to a small guard.

Table 2.1 M2 messages between the MCE and the eNodeB

eNodeB	MCE	Parameters
←	M2 SETUP RESPONSE	MCCH-related BCCH configuration PDCCH length repetition period offset subframe allocation period modulation and coding scheme
←	EnodeB CONFIGURATION UPDATE ACK	MCCH-related BCCH configuration
←	MCE CONFIGURATION UPDATE	MCCH-related BCCH configuration
←	MBMS SCHEDULING INFORMATION	PMCH Configuration List

2.5.2.3 Behavior of an MBMS-enabled UE

Such a UE is able to decode the MCCHs which are multiplexed in Type 13 system information messages. The UE through a user interface has selected which MBMS service identified by a TMGI and flow identifier it wishes to receive. The UE scans the various frames and subframes to find those containing MCCH blocks. They are decoded to search for the relevant TMGI x flow identifier. When this is the case, the MTCH subframe (the signal) and the logical channel identity can be extracted and passed to the application in the UE (e.g., TV channels).

2.5.2.4 Optimization of the MBMS channel allocation between federated groups

The logically centralized MCE will perform a re-optimization of the allocation of MBMS channels when a tactical network joins or leaves an existing federation; this is what is meant by the *C* for "coordination." The mathematically correct model is assuming that we have a total of six federated groups which have various MBMS resource requirements (if visio is used, more bandwidth is necessary than telephony PTT only or browsing only the police applications). We can assume that all eNodeB are identical and have a capacity of B which can be allocated to the various service areas.

Let the six bandwidth requirements (Mbps) be: 20,10,5,1,3,8

Let the six relative mission importance c_i of each group i be: 80,20,3, 1,5,10

Let the common capacity of B of each cell for MBMS be: 43 Mbps

Let the x_i be the proportion of the requirement i which is satisfied in the optimal solution

$$\max P = \Sigma c_i x_i, \quad 0 <= i <= N \text{ number of groups (6 here)}$$
$$\Sigma a_i x_i <= B, \quad \text{for all cells as their MBMS resources are identical}$$
$$0 <= x_i <= 1,$$

This is a simple linear programming problem and the optimum (see [0.5] for the demonstration) is computed by the centralized MCE by sorting the six ratios c_i/a_i and allocating the MBMS resources in that order until all requirements of a group are fully satisfied or the cell common maximum allocated cumulated MBMS resources are reached.

$$80/20 >= 20/10 >= 5/3 >= 10/8 >= 1/1 >= 3/5$$
$$x1 = 1 \quad x2 = 1 \quad x5 = 1 \quad x6 = 1 \quad \mathbf{x5=2/3} \quad \mathbf{x3=0}$$

In bold, one sees the groups #5 and #3 which requirements cannot be fully met and group 3 which mission importance is low cannot have any MBMS resources.

2.5.2.5 Meaning of MBMS synchronization, role of the MCE

All the eNodeBs are clock synchronized including GPS; synchronization means then that the MCCH and MTCH occupy the same subframes sent by different eNodeBs of the same MBSFN area. This also means that the "MCCH-related BCCH configuration" is the same for all eNodeBs of an MBFSN in the various M2AP messages which contain this information.

2.6 Overview of a PMR or local loop network architecture: inclusion of direct calls' support

In Figure 2.1, we have seen some of the "equipment" of a tactical network; the quote comes from the now frequent virtualized implementation on a single physical machine (see Figure 6.4 for an example). A key element is the HLR-HSS, which has in principle many standardized interfaces in two main categories: MAP and Diameter as illustrated.

2.6.1 PMR HLR-HSS capabilities and architecture

A comprehensive HLR-HSS for PMR networks and beyond are represented in Figure 2.5.

Concerning the interfaces more specifically referred in chapters of this book:

- S6m, S6t, S6n/Diameter → Chapter 10 on Internet of Things
- SLh/Diameter → Chapter 3 on Geoloc
- Zh/Diameter → Chapter 6 on MBMS
- X1, X2, X3 → Chapter 9 on Legal Interception
- PC4a → Proximity services(ProSe) also called direct calls "talkie-walkie" between two handsets (see the protocols in the next section).

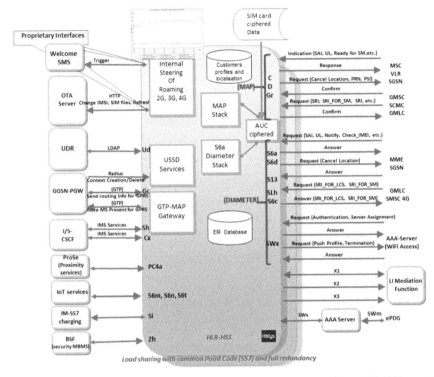

Figure 2.5   Interfaces of a comprehensive HLR-HSS for PMR, local loop 4G, IoT, as well as MNOs.

2.6.2 Proximity services (ProSe)

The HLS-HSS provides the authorization to enter direct calls groups in the 3GPP standard model (Table 2.2).

The migration from the current Tetra or P25 PMR systems to full LTE cannot be immediate as the proximity direct call service is mandatory for many users. There are ideas among some Tetra RAN manufacturers to *have a common EPC managing LTE eNodeBs and Tetra stations*. The authentication algorithms are completely different [2.12] and based on identities which are different in length and structure. That would require difficult conversions in the Tetra stations which may not worth the saving. To use a standard HLR-HSS which needs a six-digit MCC-MNC in the IMSI (the Tetra identity is six digits), a conversion ISSI(Tetra) → IMSI(4G–3G) table is required *in each Tetra station*, which means that IMSI should be individually provisioned *in each Tetra station*, which is impractical in a real situation. The only way is

Table 2.2   ProSe primitives with the HLR-HSS

PC4a Command/Diameter TS 29.344	Short Name and Direction	Explanation
Prose Subscriber Information Request	PIR ProSe Function → HSS	User-Name
Prose Subscriber Information Answer	PIA ProSe Function- ← HSS	
Prose Notify Request	PNR ProSe Function → HSS	User-Name ProSe-Permission Visited-PLMN-Id
Prose Notify Answer	PNR ProSe Function ← HSS	
ProSe Initial Location Information Request	PSR ProSe Function → HSS	User-Name
ProSe Initial Location Information Answer	PSA ProSe Function ← HSS	
Update ProSe Subscriber Data Request	UPR ProSe Function → HSS	
Update Prose Subscriber Data Answer	UPR Prose Function ← HSS	
Reset Request	RSR ProSe Function → HSS	
Reset Answer	RSA ProSe Function ← HSS	

to have the conversion table in the HLR-HSS which is not standard then, also leaving the question of different authentication algorithms Milenage and TA11–TA12 although the principle of a random challenge is the same.

References

[2.1] A. Henry-Labordère, T. Braconnier, S. Cruaux, J.P. Cohet, G. Duporche, B. Mathian, "Système de réseau PMR formé d'un réseau central RC et d'au moins un réseau local RL de téléphonie mobile," Patent FR 1751977, 2017, *Hierachical federation of several tactical networks with a central network for the global intergroup MCPTT service where the local networks HLR-HSS becomes transparent and the MCPTT is provided centrally.*

[2.2] "LTE Release 9 Technology White Paper," *tells that the maximum number of subframes usable for MBMS is 6,* available at: cdn.rhode-schwarz.com

[2.3] R. Ferrus, O. Sallent, "Mobile Broadband Communications for Public Safety," Wiley, 2015.

[2.4] M. Anis, X. Lagrange, R. Pyndiah, "Overview of evolved Multimedia Broadcast Multicast Services (eMBMS)," available at: hal.archives-ouvertes.fr/hal-01291201.

[2.5] TS 29.109 V14.0.0 (2017-05), "Digital cellular telecommunications system (Phase 2+) (GSM); Universal Mobile Telecommunications System (UMTS); LTE; Generic Authentication Architecture (GAA); Zh and Zn Interfaces based on the Diameter protocol; Stage 3," Release 14.

[2.6] 3GPP TS 23.303 v12.5.0 (2015-07), "Universal Mobile Telecommunications System (UMTS); LTE; Proximity-based services (ProSe); Stage 2," Release 12, *Specifications of the "direct mode" functions when the handsets are out of network coverage.*

[2.7] 3GPP TS 29.344 v12.3.0 (2015-03), "Universal Mobile Telecommunications System (UMTS); LTE; Proximity-services (ProSe) function to Home Subscriber Server (HSS) aspects; Stage 3," Release 12, *interface PC4a between AS ProSE and HSS which uses the ProSe primitive "Subscriber Information Retrieval Request" which returns the AVP Prose-subscription-data. AVP also in S6a.* Tetra comparison (IMSI 3GPP = 14–15 digits (Milenage algorithm), ISSI Tetra = 6 digits, (TA11–TA12 algorithms))

[2.8] ETSI EN 300 392-7 V2.2.1, "Terrestrial Trunked Radio (TETRA); Voice plus Data (V+D); Part 7: Security," September 2004.

[2.9] ETSI EN 302 109 V1.1.1, "Terrestrial Trunked Radio (TETRA); Security: Synchronization mechanism for end-to-end encryption," October 2004.

[2.10] TETRA MoU SFPG Recommendation 02 edition 4, "End-to-End Encryption," October 2004.

[2.11] TETRA MoU SFPG Recommendation 01 edition 4, "TETRA Key Distribution," February 2006.

[2.12] "The Vulnerability Analysis and Improvement of the TETRA Authentication Protocol," available at: www.icact.org/upload/2010/0423/2010 0423_Abstract_B.pdf

3

Geo-Localization of PMR Group Members and Monitoring of the Quality of Service with the ECID Method

Demaisons

- *Who are those peoples which moves will be studied?*

Blandin (the jealous location system inventor, based on a living cell extracted from the target)

- *Their names C120, C88, A37 clearly suggest that they are secret agents whose comings and goings must be controlled. In fact, I do not know any more about them than you do. I was told this is a secret subject with a penalty for high treason.*

"La parcelle Z," Jacques Spitz, 1942, J.Vigneau editor, page 33.

The story does not finish well, Blandin had taken a cell of his own young wife. He thought wrongly she was unfaithful, and kills her at the end, she was not visiting a lover but her old mother.

3.1 Operational need for a geo-localization service in PMR networks

Very much as in the military tactical networks, when the security forces are involved in a mission, they should be able to have a service which displays on their handset the position of the other group members. This is achieved by their Mission AS which receives the position of each member and broadcasts a map with all the updated positions.

For the central network management, it is important to have an *accurate monitoring of the end-to-end QoS* especially when group members are far from an eNodeB as it could mean their loss of complete service.

3.2 Localization methods in tactical networks

The usual localization methods used in PMR [3.13, 3.14] such as TETRA are "mobile initiated" and based on a specific client in each UE which transmits every few seconds the GPS position to the MCPTT. The MCPTT sends the position of each group member to each member as a map. So, control-plane localization methods or OMA SUPL (user-plane localization) are not used for this operational application. In general (Tetra or LTE), there is a proprietary application in the UE which transmits the GPS coordinates every few seconds to the MCPTT server; it is able then to refresh all UEs with the map having the display of all the UEs' positions.

Note that this requires having an offline map loaded in the local MCPTT server in case the tactical network is not connected to the Internet.

3.2.1 Enabling the LPP protocol in the UEs

However, the control-plane-based ECID method using *LPP to communicate with the UEs* is most useful to display the reference signal received power (RSRP) of the UE *in a tactical network* the features of which are configured by the PMR operator for this purpose. If an eNodeB is used at full capacity, a UE with a poor RSRP could be prioritized and get a better bitrate thanks to a PCRF. Hence, it is important that the *UEs or CPEs have the LPP protocol enabled* and that the core network has a serving mobile location center (SMLC) which is ECID capable, but the LPP protocol is transparent for the eNodeBs (not the case of LPPa).

A tactical network is a full EPC, and its MME, the features of which are selected by the PMR operator, will have the control plane interfaces Slg [3.3] and Sls [3.4] to communicate with the SMLC and GMLC with a control screen such as Figure 3.1.

3.2.2 Using SUPL as main geo-localization protocol

There is an interest in using a standard user plane method as the user can be outside the PM own coverage, and the public network used is not assumed to have the features for CP methods. On the other hand, a normal data access to the PMR applications provides the path with an SLP server using SUPL [3.7–3.11].

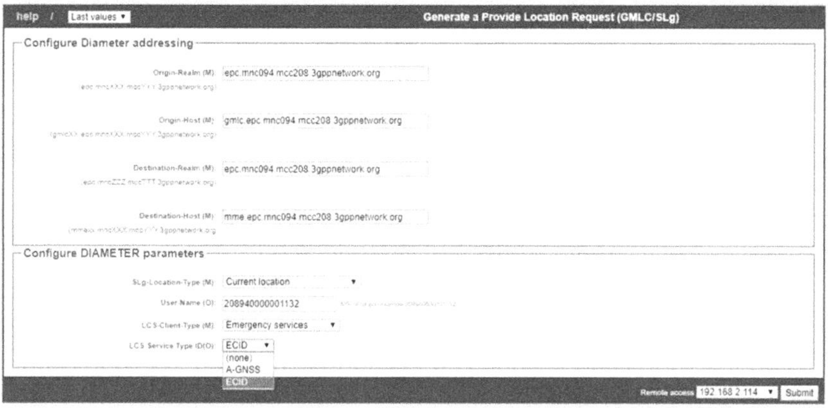

Figure 3.1 Choice of Positioning Method: ECID or GPS with the GMLC user interface.

3.3 ECID positioning method (LPP control plane) using a graphic interface

Prerequisites: the tactical network has a GMLC and an SMLC, the MME has the SLg interface with the GMLC and SLs interface with the SMLC, and all the UEs are LPP enabled.

The SLg Provide Location Request messages can be used to ask for the UE location. The positioning method can be chosen in the GMLC interface (Figure 3.2):

With the ECID choice, the GMLC asks the UE to provide signal measurements of its serving cell and neighbor cells. One of the measurements available for the serving cell (UE Rx–Tx time difference) is used to deduce the distance between the UE and the eNodeB:

As you can see the Received Signal RSRP is −59 dB which is a strong signal as the UE is about 15.9 m from the eNodeB.

3.4 Cell database for the ECID method yielding the UE received signal level

The E-CID positioning method also needs that the SMLC knows all eNodeB GPS coordinates of its network. Cell GPS coordinates can either be entered manually or extracted from an eNodeB management database. Azimuth of

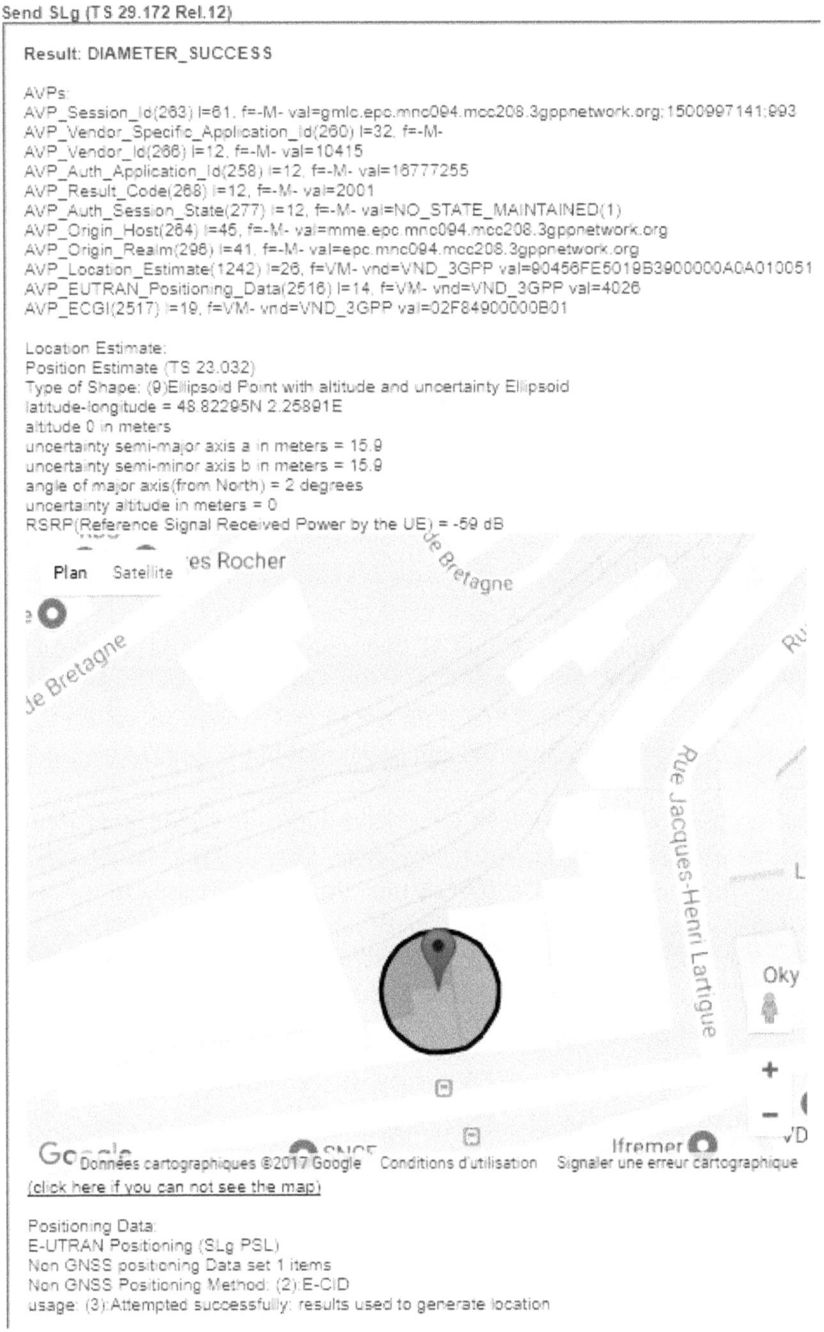
Send SLg (TS 29.172 Rel.12)
Result: DIAMETER_SUCCESS
AVPs:
AVP_Session_Id(263) l=61, f=-M- val=gmlc.epc.mnc094.mcc208.3gppnetwork.org;1500997141;993
AVP_Vendor_Specific_Application_Id(260) l=32, f=-M-
AVP_Vendor_Id(266) l=12, f=-M- val=10415
AVP_Auth_Application_Id(258) l=12, f=-M- val=16777255
AVP_Result_Code(268) l=12, f=-M- val=2001
AVP_Auth_Session_State(277) l=12, f=-M- val=NO_STATE_MAINTAINED(1)
AVP_Origin_Host(264) l=45, f=-M- val=mme.epc.mnc094.mcc208.3gppnetwork.org
AVP_Origin_Realm(296) l=41, f=-M- val=epc.mnc094.mcc208.3gppnetwork.org
AVP_Location_Estimate(1242) l=26, f=VM- vnd=VND_3GPP val=90456FE5019B3900000A0A010051
AVP_EUTRAN_Positioning_Data(2516) l=14, f=VM- vnd=VND_3GPP val=4026
AVP_ECGI(2517) l=19, f=VM- vnd=VND_3GPP val=02F84900000B01
Location Estimate:
Position Estimate (TS 23.032)
Type of Shape: (9)Ellipsoid Point with altitude and uncertainty Ellipsoid
latitude-longitude = 48.82295N 2.25891E
altitude 0 in meters
uncertainty semi-major axis a in meters = 15.9
uncertainty semi-minor axis b in meters = 15.9
angle of major axis(from North) = 2 degrees
uncertainty altitude in meters = 0
RSRP(Reference Signal Received Power by the UE) = -59 dB
Plan
Satellite
es Rocher
de Bretagne
de Bretagne
Rue Jacques-Henri Lartigue
Oky
Ifremer
Données cartographiques ©2017 Google
Conditions d'utilisation
Signaler une erreur cartographique
(click here if you can not see the map)
Positioning Data:
E-UTRAN Positioning (SLg PSL)
Non GNSS positioning Data set 1 items
Non GNSS Positioning Method: (2):E-CID
usage: (3):Attempted successfully: results used to generate location

Figure 3.2 Result of LBS positioning method ECID.

Executive Telecom(Halys)

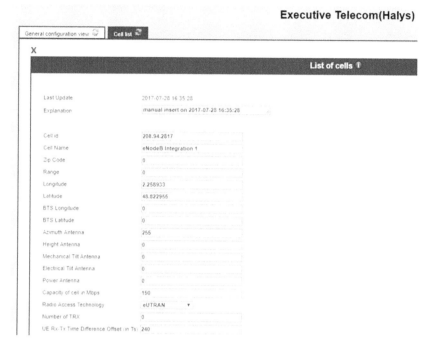

General configuration view	Cell list	

X

List of cells

Last Update	2017-07-28 16:35:28
Explanation	manual insert on 2017-07-28 16:35:28
Cell id	208.94.2817
Cell Name	eNodeB Integration 1
Zip Code	0
Range	0
Longitude	2.258933
Latitude	48.822955
BTS Longitude	0
BTS Latitude	0
Azimuth Antenna	255
Height Antenna	0
Mechanical Tilt Antenna	0
Electrical Tilt Antenna	0
Power Antenna	0
Capacity of cell in Mbps	150
Radio Access Technology	eUTRAN
Number of TRX	0
UE Rx-Tx Time Difference Offset (in Ts)	240

Figure 3.3 Cell configuration with the GPS coordinates of the eNodeB, the eNodeB delay, and the DL max bandwidth.

the antenna can also be entered for the eNodeB with a directional antenna (Figure 3.3).

As you can see the **eNodeB_delay,** the role of which is explained below, should be included in the data as it allows having the value of an offset used in the ECID method. It is specific of an eNodeB model and sw version. It can be easily found by calibrating each new model which is used by measuring the UE Rx–Tx with a UE at a few meters from the antenna.

3.5 Why not use GPS positioning method (LPP control plane)?

LPP control-plane positioning can also use the GPS positioning method to get a precise GPS position directly from the UE. The same prerequisites as for the E-CID method are needed (except that eNodeB coordinates are not needed).

Result: DIAMETER_SUCCESS

Location Estimate:
Position Estimate (TS 23.032)
Type of Shape: (9)Ellipsoid Point with altitude and uncertainty Ellipsoid
latitude-longitude= 48.82273N 2.25848E
altitude 49 in meters
uncertainty semi-major axis a in meters= 57.3
uncertainty semi-minor axis b in meters = 7.7
angle of major axis(from North) = 72 degrees
uncertainty altitude in meters = 34
confidence= 68 per cent

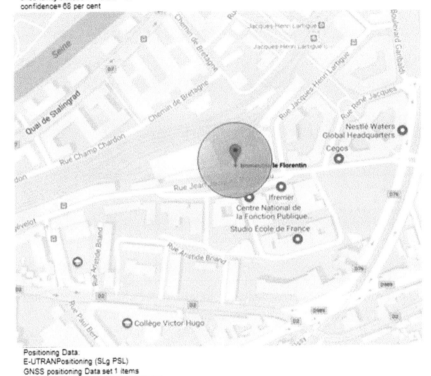

Positioning Data:
E-UTRANPositioning (SLg PSL)
GNSS positioning Data set 1 items
GNSS Positioning Method: (0): UE-Based
GNSS-ID: (0): GPS
usage: (3):Attempted successfully: results used to generate location

Figure 3.4   Result with the GPS UE-based method.

GPS also needs to be activated in the UE and the reply from a UE to a GPS
positioning request can take up to 1 or 2 minutes if A-GPS (assisted GPS, a
server helping the UE to get its initial position) is not available (Figure 3.4).

3.6 ECID method: Calculation of the physical measures from the measurements received from the UE

The distance estimation is not as trivial as many think and will be explained. The signal power is straightforward.

3.6.1 RSRP measurement → Dbm values for signal level at the UE

- In LTE, RSRP of a UE is considered excellent above -80 or -90 dBm and poor under -110 or -120 dBm. The "high-quality signal" limit as used in PLMN automatic selection procedure (3GPP TS 23.122) is set to -110 dBm (TS 36.304 ch 5.1.2.2).

3.6.2 UE Rx–Tx → distance estimate between UE and eNodeB

The LPP measurement with ECID returns the parameter "UERx–Tx time difference," measured in Ts basic timing unit. How to derive the most accurate distance measurement and what is the accuracy. Read very carefully [page 12, par. 5.2.5, 3.1] which defines UE Rx–Tx.

- *Definition "The UE Rx–Tx time difference is defined as TUE-RX – TUE-TX," where TUE-RX is the UE received timing of downlink radio frame #i from the serving cell, defined by the first detected path in time. TUE-TX is the UE transmit timing of uplink radio frame #i.*
- *The reference point for the UE Rx–Tx time difference measurement shall be the UE antenna connector.*

The specification implies that when the UE receives an LPP Position Request method ECID, it sends an uplink radio frame to the eNodeB to be echoed back.

eNodeB ← uplink radio frame #i ——- Tx←UE

- **+ eNodeB_delay** → downlink radio frame#i—- Rx>– UE

Let us note the exact distance between the eNodeB and the UE, and the eNodeB_delay, the delay for the eNodeB to echo the uplink radio frame # I. As the reference point is the UE antenna connector, it means that the UE is able to calibrate its "UE delay" and *compensates for it in estimating Tx and Rx* at the UE antenna connector. It remains the eNodeB_delay which the UE cannot estimate. We have:

$$Rx = Tx + \mathbf{d}/c \text{ (uplink)} + eNodeB\_delay + \mathbf{d}/c \text{ (downlink)}$$

Table 3.1 Coverage comparison between various eNodeBs

Manufacturer and Model	Power	Frequency	Antenna	Maximum Usable Distance	ENodeB_Delay in Ts and Meters	Date of Test
Airspan Synergy 2000 Pico	30 dBm	2,6 GHz (Band 7)	Small built-in directional	100 m	120 (1100 m)	26/7/2017
Air-Lynx	43 dBm	700 MHz (Band 28 PPDR)	Omnidirectional on an 8 m mast	800 m	20 (196 m)	1/6/2017

which gives the equation so that the SMLC in the PMR tactical network can compute the distance d:

$$d = c(\text{UE Rx} - \text{Tx} - \textbf{eNodeB\_delay})/2 \qquad (3.1)$$

One needs to have the value of eNodeB_delay, which is variable depending on the eNodeB model. In a PMR tactical network, there is a single eNodeB; it can be calibrated once for all and the eNodeB_delay entered once for all in the cell database using that model of eNodeB. The spec is clever asking each UE to calibrate its measurement so that measurements by various unknown UE models can be used.

3.6.3 Field results and coverage comparisons between various eNodeBs

The only purpose of this comparison, the frequency, the prices, the antennas, etc., being not comparable is to show that the eNodeB delay specific of a model can also be very different and that it is easy to make a calibration of the model value by making an ECID-based distance measurement when a UE is quite close to the eNodeB (Table 3.1). The EnodeB_delay corresponding to the used eNodeB is then permanently entered in the cell database.

Measurements of UE Rx–Tx time difference in E-CID positioning have an accuracy of ± 20 Ts (basic timing unit defined in TS 36.211) under normal power conditions (-50 to -115 dBm). From equation (3.1), this gives a distance accuracy of $\pm (20 \times 9.31)/2$ that is ± 93 m. Table 3.2 allows to calibrate the distance-RSRP function for the E-CID positioning.

3.6.4 Operational use and presentation of the ECID method results in PMR tactical networks

Figure 5.2 will show a very comprehensive tactical network with a virtualized architecture including SMLC and GMLC. Its supervision tool allows obtaining the received signal power of a particular UE such as Figure 3.5.

Table 3.2   Display of all the received signal power of all UEs or CPEs

Received Signal Power	Qos Parameters QCI, GBR, etc.	Current Throughput (Last Seconds)	Distance to eNodeB
−110 dBm		0.084 Mb/sec	800 m
−105 dBm		0.120 Mb/sec	750 m
−90 dBm		7.408 Mb/sec	220 m
−85 dBm		12.834 Mb/sec	450 m
−65 dBm		12.530 Mb/sec	240 m

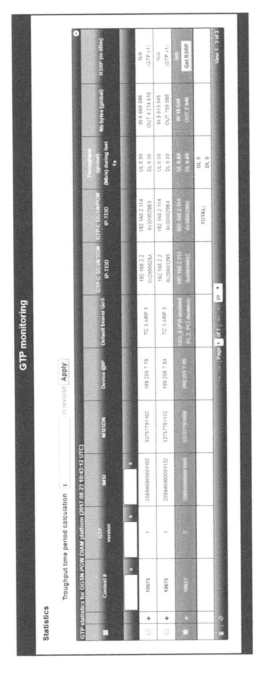

Figure 3.5   Display of all the received signal power of a given UE or CPE (get RSRP click button).

The supervision also includes a global command which obtains the RSRP *of all UE connected* and then can present an overall quality summary starting from the lowest. The GGSN-PGW contexts are used to obtain the list. Then received signal power and distance estimate are measured by ECID using the LPP protocol between the SMLC and the UEs or CPEs. The distance accuracy of a single measurement is ±10 Ts that is ±93 m (Figure 3.5).

References

3GPP Control Plane 4G

[3.1] 3GPP TS 36.214 (2017-3) V14.2.0, "LTE; Evolved Universal Terrestrial Radio Access (E-UTRA); Physical layer; Measurements," Section 5.2.5, page 12, computation of UE Rx-Tx.

[3.2] TS 36 211 V12.5.0 (2015-07), "LTE; Evolved Terrestrial Radio Access (E-UTRA), Physical channels and modulation."

[3.3] 3GPP TS 29.172 v12.5.0 (2015-04), "Digital Cellular Telecommunications system (Phase 2+); Universal Mobile Telecommunications System (UMTS); LTE; Location Services (LCS); Evolved Packet Core (EPC) LCS Protocol (ELP) between the Gateway Mobile Location Centre (GMLC) and the Mobile Management Entity (MME); **SLg Interface,**" Release 12, *Coding of PROVIDE SUBSCRIBER LOCATION 4G.*

[3.4] 3GPP TS 29.171 v12.1.0 (2014-10), "Digital Cellular Telecommunications system (Phase 2+); Universal Mobile Telecommunications System (UMTS); LTE; Location Services (LCS); LCS Application Protocol (LCS-AP) between the Mobile Management Entity (MME) and Evolved Serving Mobile Location Centre (E-SMLC); **SLs interface,**" Release 12, Equivalent 4G du BSSAP-LE et du PCAP.

[3.5] 3GPP TS 36.355 v12.4.0 (2015-04), "LTE; Evolved Universal Terrestrial Radio Access (E-UTRA; LTE positioning Protocol (**LPP**)," Rel12, *LPP is the protocol between E-SMLC and UE.*

[3.6] 3GPP TS 29.173 v12.3.0 (2015-04), "Digital Cellular Telecommunications system (Phase 2+); Universal Mobile Telecommunications System (UMTS); LTE; Location Services (LCS); "Diameter Based **SLh** interface for Control Plane LCS" Release 12, *Equivalent 4G of MAP SRI_FOR_LCS.*

OMA SUPL (Secured User Plane Location)

[3.7] OMA-TS-ULP-V2_0_1-20121205-A, "UserPlane Location Protocol," December, 2012.

[3.8] OMA-TS-MLP-V3_2-20051124C, "Mobile Location Protocol."

[3.9] OMA-TS-ULP-V3_0-20140916-C, "User Plane Location Protocol," *Network initiated geoloc for emergency services and address verification of the H-SLP.*

[3.10] OMA-AD-SUPL-V2_0-20120417-A, "Secured User Plane Location Protocol," *Security in the H-SLP authentication.*

[3.11] OMA-WAP-TS-ProvSC-V1_1-20080226-C, *"Provisioning of smart cards with the configuration PKCS15 configuration."*

TETRA LIP (Location Information Protocol)

[3.12] ETSI TS 100 392-18-1 V1.7.1 (2015-03), "Terrestrial Trunked Radio (TETRA); Voice plus Data (V+D) and Direct Mode Operation (DMO); Part 18: Air interface optimized applications; Sub-part 1: Location Information Protocol (LIP)."

[3.13] ETSI TS 100 392-18-4 V1.2.1 (2015-07), "Terrestrial Trunked Radio (TETRA); Voice plus Data (V+D) and Direct Mode Operation (DMO); Part 18: Air interface optimized applications; Sub-part 4: Net Assist Protocol 2 (NAP2)," *Equivalent TETRA of SUPL for GPS assistance.*

TETRAPOL (Cassidian-Airbus TETRA varian)

[3.14] INFOCERT (GT 399 Antarès), "Document de Specification Interface Inter SGP," July, 2014.

[3.15] Cassidian, "Mobile Data Gateway (MDG) Application Interface," Rev 10 26/10/2012, *Interface with the geo-localisation server TETRA ou TETRAPOL based on OMA-MPL transported by HTTP GET and POST.*

[3.16] EADS, "AVL Server: External Interface Description," MC9/SYS/DD/00291/03/13/EN, September, 2008. *UDP based interface for MCPTT with the Automatic Vehicle Location server.*

4

Choice of the SIM Card Type for PMR or M2M Networks and Automatic Profile Switching Possibilities

During the Renaissance and up to now they were only three doctorates: theology, law, and medicine. This why famous scientists had a doctorate in medicine and why the doctorate in science is created.
Décret of the 17 march 1808 (France)

4.1 Classical UICC, eUICC M2M, or eUICC "consumer" SIM cards

4.1.1 Usage difference

Should the PMR or Internet of Things (IoT; also called M2M machine-to-machine) networks use:

- the classical USIM cards with multi-IMSI Applets (they may have a single security domain or one per IMSI) for Ki, OPC end the milenage personalization parameters
- or the eUICC type SIMs (several security domains, one for each IMSI) designed for M2M?

The two types of cards support the 3GPP TS 23.048 standard of 4.3.1.3 for the remote installation and delete of Applets via I/O or OTA.

The IMSI (standard EF file) can be updated through OTA. OTA 23.048 security keys can be changed using the PUT KEY, if the cards are created with the proper update rights. The GET DATA command can be used with vendor-specific TAGs to recover specific applet data.

PMRs have the additional operational requirement, but not least: *IMSI switch* (automatic or manual) when the device loses the network coverage.

There are two influences which promote eUICC based on commercial reasons:

- *SIM card vendors* will push the eUICC choice because it keeps a dependence with their SM system for an IMSI switch (and security domain) for a long time contract.
- *MNOs candidating to provide the resilience* networks will also push the eUICC, so resilience entails a complete switch to their own security domain and they can use all their existing HLR-HSS setup considering the PMR subscribers as ordinary subscribers.

There is also a third choice to discuss: eUICC "Consumer" where switch of IMSI has to be controlled by an application in the device. This application offers the possibility to have a manual switch from one profile to another, but *making it automatic with a LAP with eUICC is not demonstrated. An SM connection* would have only visible PLMNid(s) as selection criterion in any case.

Classical USIM cards provide a largely proven and autonomous way of switching the IMSI in case of a loss of coverage with a multi-IMSI Applet. We see below that security domain updates are possible by OTA with some SIM card vendors.

eUICC M2M in different formats including soldered was designed for a typical automotive case. It equips vehicles for monitoring which leave the factory (example BMW) with a German mobile subscription (for example, Vodafone D2) which has roaming agreements in most of the countries. For example, when the vehicle arrives in China, it can have a data connection with BMW monitoring center which has an SM and is able to remotely create a new (for example, China Mobile) profile with an IMSI (like a new SIM card). Then remotely the China Mobile profile is activated by the SM, and the German initial profile is never used again. The M2M is not designed for quick and autonomous (it needs a connection to the SM) switch. The standard GSMa SM architecture is in Figure 4.1.

Note that eUICC M2M may automatically switch to a (single) emergency account, but this is not a proper mechanism for PMR resilience.

4.1.2 Difference of logical structure between UICC and eUICC

4.1.2.1 eUICC
The eUICC can be considered as separate SIMs ("Profiles") with *no common files*. That makes an automatic profile switch *by the SIM* impossible, although it is possible:

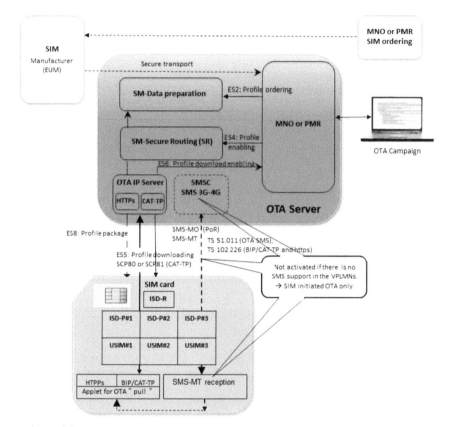

Figure 4.1 Remote provisioning of a SIM card, eUICC (embedded UICC), or UICC.

- from a connected subscription manager
- from a local profile assistance in the device which sends switching commands to the eUICC.

4.1.2.2 UICC

On the contrary, UICC with multi-IMSI applet is one standard SIM with standard files (IMSI, preferences, and SMS parameters) and security keys which are copied from those associated with a particular IMSI when an IMSI is selected. To make the change of security keys possible, they are not in protected files. At least the following SIM files are switched in most vendors' implementation:

- EFspn (service provider name) which shows on the handset
- EFimsi, the IMSI of course

- EFacc, access control class (many PMR uses 11 = PLMN)
- EFsmsp, SMS parameters (the SMSC global title)
- EFad, administrative data which especially contain the length of MNC in the IMSI
- EFspdi, service provider display information.

But if the size of the card is sufficient (256K), any files can be included in the multi-IMSI applet and switched. There is a logic that the subscriber is the same and it is illogical to switch the EFli (language indication) file for example.

4.1.2.3 Recent file additions for all card types [4.2]
Note the new structure files BER-TLV (added to well-known transparent, linear fixed, cyclic) to store objects and used, for example for EFmcptt_config. There are also parameter files for ProSE, MBMS, and multimedia which are reserved while the usage is not well specified.

4.2 Remote provisioning system for eUICC (M2M and consumer)

Figure 4.1 shows the interfaces between the eUICC and the subscription manager (SM; off-card interfaces) in the remote provisioning. The eUICC contains:

- ISD-R (issuer security domain root)
- USIM (multiple instances possible)
- Profiles (multiple instances possible)
- multi-IMSI Applet files for different security domains' IMSIs
- an important Applet for "OTA IP Pull," either BIP-CATTP or https or both.

The data transport protocol is SCP 80/81 (CAT-TP), SCP 80 being the SMS bearer and security [4.5]. SCP03 (TLS) is used for the ciphering and the data protection between the service and the security domain in the eUICC.

Subscription manager secure routing (SM-SR), SM-DP (data preparation), and MNO belong to the off-card domain off-card. The rest belongs to the eUICC.

In the GSMa sense, only the card manufacturer and the MNO have the transport keys and the security keys.

Once the eUICC has been switched to a given profile (a UICC and its IMSI), the OTA update of the various SIM files is exactly the same as an ordinary UICC with its single security domain.

After a reboot (power ON/OFF) of the eUICC, the security domain remains the same. For UICC, there is a single one common to all IMSIs, so it remains the same of course.

4.2.1 Explanation of the remote provisioning figure

Figure 4.1 shows an optional SMSC 3G-4G *which will not be activated* if the tactical network PMR have no SMS capability in their MME, or if the users cannot be under the coverage of a public network which has it (at least 4G). In that case, the SIMs *must have the Applet for OTA "pull,"* that "is card initiated OTA," which either connects periodically to the OTA server or provides a menu displayed in the terminal so that users may initiate an OTA over IP update of their SIM.

If SMSs are available, it could be used for only legacy SMS OTA (commands TS 51.011 [4.6]). It is also used to send SMS-MT channels' establishment commands TS 102.226 [4.10] with the parameter P2 (see Section 4.5.2.2) set to trigger either a UDP (BIP-CATTP) or TCP (https) connection to the OTA server. This is why there is an arrow between the SMS-MT reception function in the SIM card and the Applet for OTA "pull" to show how it works: an SMS-MT is received by the SIM and sent to the Applet.

4.3 eUICC and UICC profile switching methods

Table 4.1 shows the different profile switching methods depending on the SIM card manufacturers' choices:

4.3.1 Add IMSI with its own security domain in UICC by OTA

The UICC and the multi-IMSI applet must support this functionality.

4.3.1.1 Logical organization of a multi-security domain UICC SIM card

During the IMSI switch, the applet takes the security keys (Ki, OPc, and milenage parameters) of the new IMSI from the multi-IMSI files and copies them in the "normal" place.

When there is change of IMSI, the security keys of the IMSI selected by the multi-IMSI Applet are copied in the normal Ki, OPc, and milenage files *which are not locked for the Applet which has the rights for writing* these files. This is the characteristic of UICCs designed for multiple security domains.

Table 4.1   Switching methods depending on the SIM card type

SIM Card Type	Security Domain HLR-HSS Ki, Opc, Milenage	Security OTA 23.048 KiC, KiD, KiK	Profile Switching Mode	Switching Criteria	Adding a Profile-IMSI	Example
eUICC M2M (e.g., vehicles) no need for a rapid change of security domains.	Multiple, one per profile	Multiple, one per profile	1) *by the SM platform* to which there must be a connection 2) *manual* switch by an application to a single "emergency profile" 3) *automatic* mode possible by a special application *in the "device"*	The same parameters available as for the eUICC "Consumer" below Decision by the SM platform	With the SM only	Telit and Sierra wireless modems, which may change profile when the modem goes to another country
eUICC consumer	Multiple, one per profile	Multiple, one per profile	(Local profile assistance) application in the handset.	For an automatic profile switch (change of IMSI), the available measurements for the application *may(**)* be: PLMNid (MCC-MNC) EARFN (allows to recognize the PPDR frequency allocated to the PMR network, RSRP (received signal power by the handset), RSRQ (received signal quality which is the signal/noise ratio SNR; interesting to distinguish if the PPDR is sideband of a commercial band such as 28.39/209	With the SM only	France defense sector. **Non-standard terminal** with two modems.

UICC	*Multiple* and *single* security set of keys possible (specified at order)	Multiple or single set of keys for all IMSIs	Automatics by a multi-IMSI Applet or manual switch (application menu in the SIM) working with any terminal	Simple: when there is no network connection possible with the active IMSI for all visible networks, the Applet selects the next IMSI and reinitializes a network selection by sending a REFRESH command to the terminal	By OTA changing the Applet multi-IMSI, see details in 4.3.1	Astrid (Belgium), different security domains for each IMSI. **Standard terminal** Time to switch: – a few seconds in the case of a coverage loss because a new network selection is started in this case immediately by the terminal. *Example: tactical network in deep indoor profound, the handset goes outside and pick up a public MNO most quickly.* – 3 minutes in average(*) to return to the PMR network once the coverage is available. During this time except a few seconds interruption, the service is provided by the MNO then by the PMR network.

*In the SIM, the parameter EFhpplmn (higher priority PLMN search period) is set to the minimum value: 1, i.e., 6 min. The minimum value to start a reselection is 2 min from point b) in Section 4.4.1.2, the average interval between 2 periodical search is then 4 min (2 + (6−2)/2) with a standard modem. This interval can be reduced for special handsets (see Section 4.4). MCC-MNC of the PMR network. This interval can be reduced for special handsets (see Section 4.4). The EFefhplmn (Equivalent HPLMN) = MCC-MNC of the PMR network.

**For the eUICC "customer" and switch by LPA application in the handset, it is not certain (depends on the terminal and the chipset) that the RSRP and RSRQ will be available to the application. Hence, this automatic switch using an LPA needs development for a particular handset.

The Applet has the same rights as the local card tools provided by the SIM card manufacturers which use a PC and a SIM card reader/writer and can write any file including Ki, Opc, and milenage. These card tools may be used to prototype a SIM card (Figure 4.2).

EFKeyKi are OTAable (e.g., Manufacturer I), not OTAble (e.g., Manufacturer J) not readable of modifiable by the local "card tool if authorized (should not because a stolen card could be cloned).

The different multi-IMSI applets tested have almost the same architecture. They have their own directory with the configuration files, the elementary files (see Section 4.1.2.2) to copy in the 3GPP TS 31.102 structure, and specific elementary files for the security keys Ki, OPc, and milenage parameters.

The structure of the multi-IMSI directory can be different from one vendor to another. Some vendors (Manufacturer J) have only one directory with big "transparent" elementary files where the file size is N x standard EF size (N is the number of IMSIs supported by the applet). Others have separate files for each domain (Manufacturer I), these are details, and the organization principle is the same.

Example: Standard EFimsi size is 9 bytes and a multi-IMSI applet is designed to manage 10 IMSIs.
$\rightarrow$ multi-IMSI EFimsi size is 9 x 10 = 90 bytes.
bytes from 1 to 9 = IMSI 1
bytes from 10 to 18 = IMSI 2
. . .

Other vendors have a different approach: the multi-IMSI directory contains several subdirectories corresponding to the IMSIs.

Note: For obvious security reasons, the elementary files where the security keys are stored must be not readable either by OTA or locally with a card reader. We can only modify them.

4.3.1.2 Add a new IMSI with its own security domain

Be careful: If the multi-IMSI applet is strictly designed to manage two IMSIs for example (multi-IMSI files' size = 2 x standard EF size or main multi-IMSI directory has only two subdirectories), the method below to add a new IMSI is not applicable. *You must order new SIM cards with BAP additional cost.*

To add a new IMSI, files listed in Section 4.1.2.2 plus security key files must be configured. It can be done with standard commands [4.5] SELECT and UPDATE BINARY/UPDATE RECORD depending on the file structure (transparent or linear fixed/cyclic).

	Ordinary SIM files	Multi-imsi Applet Files			Common
		Profile # 1	Profile # 2	Profile # 3	
31.102 common files for all IMSIs:	31.102 IMSI specific files — Ki, Opc, Milenage	IMSI #1 Ki, Opc, Milenage	IMSI #2 Ki, Opc, Milenage	IMSI #3 Ki, Opc, Milenage	-OTA Keys 23.048 -BIP/CAT-TP config(proprietary) -https config (proprietary)
Language Address Book Preference etc..	EFimsi EFspn EFsmsp EFloci EFpsloci EFacc EFad EFspdi	31.102 IMSI specific files EFimsi EFspn EFsmsp EFloci EFpsloci EFacc EFad EFspdi	31.102 IMSI specific files EFimsi EFspn EFsmsp EFloci EFpsloci EFacc EFad EFspdi	31.102 IMSI specific files #3 EFimsi#3 EFspn EFsmsp EFloci EFpsloci EFacc EFad EFspdi	-OTA 23.048 KiC, KiD, KIK -3F00/6F55 (Manufacturer J) for BIP-CAT-TP
	EFKeyKi	EFKeyK#1	EFKeyK#2	EFKeyK#3	

EFKeyKi are OTAable (e.g. Manufacturer I), not OTAble (e.g. Manufacturer J), not readable of modifiable by the local "card tool" because, if authorised, a stolen card could be cloned.

Figure 4.2 Logical structure of a multi-security domain UICC card and switch of files when a change of security domain is performed.

Note: The "normal" place at the left of Figure 4.2 where the security keys are stored *is not accessible by OTA*.

Conclusion

PMR

UICC cards *from certain card vendors*, with a multi-IMSI applet, provide the possibility of distinct security domains and can be configured by OTA. They meet all the requirements of PMR (including automatic resilience *without the dependence of a LAP application in the terminal or a connection to a subscriber management system)* and M2M networks. *The choice of the SIM card provider is important.*

They can be managed by a large number of OTA systems using push (SMS) or pull IP (BIP and CAT-TP or http(s)) with the well-known SIM file update commands [4.5].

M2M and IoT

For some vendors of UICC cards, the security domain keys are OTAble. They meet then the requirement of M2M or IoT operators which need to be sure that they can change the security domain (change of operator for commercial reasons) of their SIMs without replacing them.

MVNO going Full MVNO

These operators have a significant number of customers using the SIM of an MNO, and to become Full MVNO with their own IMSI, they will need in any case to change their SIMs. The UICC *with two security domains* or more *and OTAble Ki, Opc, and milenage* multi-IMSI Applet files is an excellent choice; they are ordered by the MNO with the same security keys Ki, Opc, and milenage as the SIMs they replace. When the Full MVNO is ready with its core network (including own HMR-HSS), he executes an OTA campaign to add its own security domain to all the new SIM cards. He may decide then to deactivate the MNO IMSI security domain, if he does not want to use the MNO as a roaming sponsor.

eUICC type SIM cards

They are proposed by card vendors for M2M applications or Full MVNOs because of their possibility with a SM system to add security domains with new IMSIs after their initial supply. This can also be done with UICC type cards, with the additional possibility of automatic autonomous IMSI switch.

Name of the Applet Management Commands TS 23.048 [4.5] or ETSI 102 226 [4.10] *(uses the TAR= Remote Applet Manager = 000000)*	Role, see [4.5]
DELETE	
INSTALL (load) LOAD (multiple inter-mediate) <————| -----------------------------| LOAD (final)	9.1.1 Package Loading The Package loading process allows the MNO to load new packages (**Applet consisting of .jar files**) onto the UICC. Depending on the size, many SMSs (OTA by SMS) might be used for the package loading and hence the interest of the IP-based PULL mode instead of SMS. One can LOAD new Applets, but cannot read them!
INSTALL (install)	9.1.2 Applet Installation The applet installation process allows the MNO to install a new application on the UICC. The installation may be only performed if the corresponding packages have already been loaded onto the card.
GET DATA (Executive Telecom UICC 256K) Payload example (<<extended card resources information>>): 001A15122115150000000000000000800 80CAFF21 Counter+paddiing counter: 000000000800 Secured payloaf: 80CAFF21 (see [4.10, page 32]) Other secured payloaf: **80CAFF20**	Reads a Data Object ISO/IEC 7816/4, response by **PoR:** *Number of installed applets: 17* *Free non-volatile memory EEPROM: 184576* *Free volatile memory: 15834* *Free E2PROM: 32767* *Number of installed applets: 17*
STORE DATA	Writes Data Object
PUT KEY for AES PUT KEY for DES	Updates KiC, Kid, KiK, the OTA TS 23.048 security keys (if the file is unlocked).
GET STATUS	
SET STATUS	

4.3.1.3 Summary of the applet management commands

These commands are used if one wanted to add or replace completely a package (applet); this is useful when there are many cards distributed.

4.3.2 Updating the OTA security keys KiC and Kid in multi-IMSI UICC cards

In two examples of Manufacturer I and Manufacturer J, the PUT KEY for DES is used to update the single set (for all IMSIs) of OTA security keys. It makes sense as a single OTA system will be used for the updates of all IMSIs. This is a difference with eUICC which has a complete separation between the IMSI security domains including the OTA TS 23.048 security keys.

4.4 Is it possible to reduce the automatic network switching time VPLMN→ HPLMN?

4.4.1 The TS 23.122 3GPP standard

It describes the initial selection and the periodical attempt to return to the PMR network coverage. We enclose an extract of this standard to avoid any ambiguity.

4.4.1.1 Automatic network selection mode procedure (Section 4.4.3.1.1 of [4.4])

The MS selects and attempts registration on other PLMN/access technology combinations, if available and allowable, in the following order:

   (i) either the HPLMN (if the EHPLMN list is not present or is empty) or the highest priority EHPLMN that is available (if the EHPLMN list is present);
  (ii) each PLMN/access technology combination in the "user-controlled PLMN selector with access technology" data file in the SIM (in priority order);
 (iii) each PLMN/access technology combination in the "operator-controlled PLMN selector with access technology" data file in the SIM (in priority order);
  (iv) other PLMN/access technology combinations with received high-quality signal in random order;
   (v) other PLMN/access technology combinations in the order of decreasing signal quality.

4.4.1.2 (In VPLMN) automatic and manual network selection modes (Section 4.4.3.3.1 of [4.4])

If the MS is in a VPLMN, the MS shall periodically attempt to obtain service on its HPLMN (if the EHPLMN list is not present or is empty) or one of its EHPLMNs (if the EHPLMN list is present) or a higher priority PLMN/access technology combinations listed in the "user-controlled PLMN selector" or "operator-controlled PLMN selector" by scanning in accordance with the requirements that are applicable to (i)–(iii) as defined in the automatic network selection mode in subclause 4.4.3.1.1. In the case that the mobile has a stored "Equivalent PLMNs" list, the mobile shall select only a PLMN if it is of a higher priority than those of the same country as the current serving PLMN which are stored in the "Equivalent PLMNs" list. For this purpose, a value of timer T may be stored in the SIM. The interpretation of the stored value depends on the radio capabilities supported by the MS:

- For an MS that does not only support any of the following or a combination of EC-GSM-IoT or Category M1 or Category NB1 (as defined in 3GPP TS 36.306 [4.19]): T is either in the range 6 min to 8 h in 6-min steps or it indicates that no periodic attempts shall be made. If no value for T is stored in the SIM, a default value of 60 min is used for T.
- For an MS that supports only any of the following or a combination of EC-GSM-IoT or Category M1 or Category NB1 (as defined in 3GPP TS 36.306 [4.19]): T is either in the range 2–240 h, using 2-h steps from 2 to 80 h and 4-h steps from 84 to 240 h, or it indicates that no periodic attempts shall be made. If no value for T is stored in the SIM, a default value of 72 h is used.

If the MS is configured with the MinimumPeriodicSearchTimer as specified in 3GPP TS 24.368 or 3GPP TS 31.102 [4.2], the MS shall not use a value for T that is less than the MinimumPeriodicSearchTimer. If the value stored in the SIM, or the default value for T (when no value is stored in the SIM) is less than the MinimumPeriodicSearchTimer, then T shall be set to the MinimumPeriodicSearchTimer.

The MS does not stop timer T, as described in 3GPP TS 24.008 and 3GPP TS 24.301, when it activates power saving mode (PSM; see 3GPP TS 23.682).

The MS can be configured for fast first higher priority PLMN search as specified in 3GPP TS 31.102 [4.2] or 3GPP TS 24.368. Fast first higher priority PLMN search is enabled if the corresponding configuration parameter is present and set to enabled. Otherwise, fast first higher priority PLMN search is disabled.

The attempts to access the HPLMN or an EHPLMN or higher priority PLMN shall be as specified below:

(a) The periodic attempts shall be performed only in automatic mode when the MS is roaming, and not while the MS is attached for emergency bearer services or has a PDN connection for emergency bearer services.

(b) The MS shall make the first attempt after a period *of at least 2 min and at most T minutes:*

- only after switch on if fast first higher priority PLMN search is disabled; or
- after switch on or upon selecting a VPLMN if fast first higher priority PLMN search is enabled.

(c) The MS shall make the following attempts if the MS is on the VPLMN at time T after the last attempt.

(d) Periodic attempts shall be performed only by the MS while in idle mode.

(d1) Periodic attempts may be postponed while the MS is in PSM (see 3GPP TS 23.682).

(e) If the HPLMN (if the EHPLMN list is not present or is empty) or an EHPLMN (if the list is present) or a higher priority PLMN is not found, the MS shall remain on the VPLMN.

(f) In steps (i)–(iii) of subclause 4.4.3.1.1, the MS shall limit its attempts to access higher priority PLMN/access technology combinations to PLMN/access technology combinations of the same country as the current serving VPLMN, as defined in Annex B.

(g) Only the priority levels of equivalent PLMNs of the same country as the current serving VPLMN, as defined in Annex B, shall be taken into account to compare with the priority level of a selected PLMN.

(h) If the PLMN of the highest priority PLMN/access technology combination available is the current VPLMN, or one of the PLMNs in the "equivalent PLMNs" list, the MS shall remain on the current PLMN/access technology combination.

4.4.1.3 Reducing the timer T

The modem accesses the T parameter of the SIM without going through the OS.

See [4.4]. In a terminal, *this standard is implemented in the "modem's firmware"* provided by the chipset manufacturer, such as Qualcomm, Mediatek, etc., *not in the OS* (Android and iOS). In order to modify this for a particular terminal, one needs to have the firmware's source code and have

Table 4.2 Possibility of adjustments of the T timer in some handsets

PMR Handset Type	Manufacturer	Model	Chipset Manufacturer	Chipset Model	Technical Possibility
Top of the range	CrossCall	Trekker x4	Qualcomm	Snapdragon 660	Yes
Middle range	Alcatel (TCL)	IdoI5S	Mediatek	MT6753	Probable, needs study

the tool to re-flash the memory. Such a non-standard change could not be available for commercial terminals. For example, there are the possibilities given by a major supplier of secured phones (Atos-Bull) which masters these modifications.

The firmware's modification is to ignore the value of in the SIM for EFhpplmn in "deci-hours," hence a minimum of 6 min, and replace it by a hard-coded or application set value (the unit can be 1 min or less).

However, it must be understood that the network selection time T is correlated to the broadcast of the signal quality from neighboring cells and 6 min is a coherent value for that. A smaller T value will also necessitate to search only for the 4G technology and it is unlikely that T can be less than 2 min from point b) in Section 4.4.1.2 in italic.

The answer is likely to be "yes" then but with a non-standard handset especially developed to allow a shorter search period.

4.5 OTA provisioning of the SIM: "Card initiated OTA SIM with IP" or "network initiated" using SMS

4.5.1 OTA SIM over IP

The use of IP protocols, while avoiding to have SMS in the network, allows to speed up remote SIM administration, most notably the download of applets (SIM Toolkits) of several kilobytes, and to have operational applications like address books' backup.

OTA SIM over IP is available with the BIP-CATTP or http(s) protocols. BIP-CATTP is more general as available in most 3G cards, while http(s) is appeared in the LTE-capable cards.

GSMa [4.1] describes the different user cases for OTA:

- legacy network initiated
- card initiated.

4.5.1.1 Legacy network initiated

This case needs SMS:

- SMS-MT to
    - select a file and read or update its content
    - trigger the open channel BIP/CAT-TP or https in order to establish an IP connection between the OTA server and the SIM card.
- SMS-MO from the card named proof of receipt (PoR) which gives the status of the command execution and the file content if a "read" command (READ BINARY or READ RECORD) has been sent.

4.5.1.2 Card initiated

This case does not need SMS at all. The card regularly connects (periodicity parameter in the SIM) to an OTA server to check if updates are available. There are two "IP methods":

1. BIP/CAT-TP [4.1]
2. http(s).

The periodic connections are possible because the SIM card has an applet "polling applet" which is in charge to establish the BIP/CAT-TP channel or the http(s) connection.

Card initiated mode offers to MNOs the possibility to provide to their subscribers specific applications like address book backup. When the subscriber selects the application, the SIM card connects to the OTA server using a BIP/CAT-TP or http(s) link and sends the address book content in order to be saved or restored.

These two different provisioning methods *are common to eUICC and UICC cards*.

4.5.2 Card initiated mode with a data connection to the OTA IP server

4.5.2.1 BIP/CAT-TP

This uses either the protocol BIP/CAT-TP (SCP 80/81) or https for all card types UICC or eUICC (M2M or "consumer"). The SIM must have the proprietary *IP OTA files of the manufacturer* filled (channel to the SM server and IP of the server). If empty, the card initiated method is not possible and the OTA server must send an SMS-MT containing OPEN CHANNEL parameters but no automatic card initiated OTA is possible.

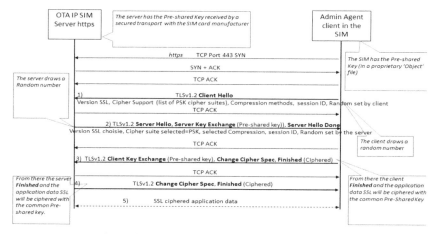

Figure 4.3   Pre-shared key (PKS) TLS handshake.

Bearer-independent protocol (BIP) concerns the link between the SIM card and the terminal to access the IP network with the adequate bearer. CAT-TP is an end-to-end connection between the SIM and the OTA IP server. To open a BIP/CAT-TP channel, there are 2 methods:

- OTA server initiated which sends a Push SM-MT coded with TS 23.048 commands [4.5] containing the OPEN CHANNEL command to request the SIM card to connect to the OTA IP server.
- SIM card initiated, no need for SMS, the card must have an Applet which periodically connects to the OTA IP server, to check eventual updates.

The BIP/CAT-TP configuration is proprietary.

4.5.2.2 OTA over https

This connection mode for OTA with IP uses the well-known protocol https. It has the same mode as BIP/CAT-TP:

- card initiated with an applet to periodically connect to the OTA IP server
- OTA server initiated with an SMS push coded with TS 102 226 [4.10]. The P2 parameter of the push command is 3 meaning "Request for TCP connection" in the case of an http(s) connection. For a BIP/CAT-TP connection, it would be 1 "Request for BIP channel opening" and then 2 for "Request for CAT-TP link establishment" which can be sent in the same SMS-MT.

Card Manufacturer	**OTA server**	**SIM card**

(computes PSK from Master Key and ICCD

with a Diversification Algorirthm)

Pre-shared Key---→ Secure Transport Key (Figure [4.1]) ---→ Pre-shared Key

←----------https TLSv1.2---------Pre-shared Key

Figure 4.4   Secured transport of PSKs between the card manufacturer and the OTA IP server.

A) Secured transport of the keys between the SIM card manufacturer and the OTA server

The "master key" has an equivalent role to the OP key used to compute OPc from the Ki (role plaid by the ICCID). This "pre-shared key" (PSK) is securely exchanged between the OTA IP platform provider and the SIM card manufacturer; it had been written as an "object" (not a standard file) in the SIM card. When an https/TLS OTA session is started, the PSK key which is used is computed with the ICCID using a proprietary "diversification algorithm" as it improves the security in the simple PSK procedure 2 which does not use certificates [4.14].

B) Security HTTPS (see Figure 4.3 with PSK)

The security for https is based on the RFC 4279 [4.14] PSK which is a shared secret between the SIM card client and the OTA server. This PSK is *stored in a SIM manufacturer's proprietary file* and is derived *from a master key for the operator and a security enhancement PIN which is the ICCID* of the SIM with the diversification algorithm. *It is then an individual PSK key authentifying a given SIM for the OTA server.*

Note that the *PSK https used for OTA SIM does not use certificates which would need a public key infrastructure (PKI) and* it is then not the same https implementation as used for secured payments between a browser and a payment site [4.16] which is based on a private certificate in the server and the verification of its public certificate by the client.

Hence, the "Certificate files" of TS 31.102 (EForpk, EFarpk, EFtprpk, and EFtkcdf) *are not used*; they are for the mobile execution environment (MeXE) [4.15] which is not used in the https OTA SIM implementation. However, this does not exclude to use MeXE-enabled devices for the subsequent secured execution of services between the UE and a server.

To select the PSK TLS V1.2 mode, the OTA IP SIM server will choose one of the two cipher suites:

- TLS_ECDHE_PSK_WITH_AES256_CBC_SHA384
- TLS_ECDHE_PSK_WITH_AES128_CBC_SHA256

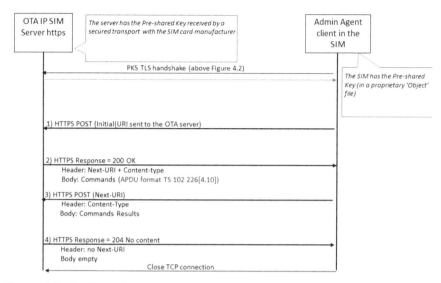

Figure 4.5 Sending of an APDU command by the OTA server to the SIM using HTTPs/TLS with PSKs.

[4.18] from the French security agency ANSI that contains all the practical details to configure an Apache or NGINX web server.

C) Secured https session to exchange secured OTA data (Figure 4.5)

On the right, the "Admin Agent" is a function of the SIM card and the OTA SIM server on the left performs the authentication of the SIM. The OTA SIM server receives the ICCID and the PSK in the PKS TLS handshake message sent by the SIM and validates the connection. Several PSKs could be stored in the card, for example, eUICCs could have different keys for each IMSI profile. The PSKs can be modified using the PUT KEY command.

You see in Figure 4.6 how the APDU payload command is coded to update the IMSI after a command which selects the directory containing the IMSI.

From the overall HTTP(s) exchange at the initiative of the SIM, the three Figures 4.3–4.5 fully practically explain how the OTA SIM using HTTP(s) works; it can be used to develop an OTA server over IP. One has to configure the server to use PSKs and not certificates for the authentication as this is the usual https security method in OTA SIM.

22 0F 00 06 00 00 0A 09 31 01 10 12 32 54 76 98 F0	C-APDU UPDATE BINARY (update of the IMSI)			
	22	C-APDU Tag (TS 100 201 [4.17])		
	0F	Length		
		Data for UPDATE BINARY		
		00	CLA value	
			see TS 102.221 [4.17] for details for the coding of Class	
		06 (UPDATE BINARY)	INS value	
			see TS 102.221 [4.17] for details for the command	
		00	P1 value	
		00	Offset low	
		0A	Length of data	
		09 31 01 10 12 32 54 76 98 F0	Data according to TS 31.102 [4.2]	
			09	Length of IMSI
			31 01 10 12 32 54 76 98 F0	1234567890

Figure 4.6   Example of expanded remote command TS 102 226 structure over HTTP(s) to update the IMSI.

4.5.3 Network initiated SMS triggering of a SIM IP connection (BIP/CAT-TP or https) to the OTA server

This *procedure is the same* to trigger a BIP/CAT-TP connection or an https connection using SMS. Appendix 11.1 (example executed with 4G SMS, the OTA server has an SMSC with the IP_SM_GW in Figure 4.1) gives the full SMS parameters details (the same for both cases) and the payload in the case of BIP/CAT-TP with the opening of the BIP and CAT-TP channels. 3G SMS or 4G SMS may be used if available

The payload is described in ETSI TS 102 223 [4.21] and the security of the trigger sending by SMS in TS 23.048 [4.5].

The SMS method is very convenient to test the SMS even if pure OTA IP service is desired, as *most basic cards have the BIP CAT-TP client*, starting in 2009 [4.21], the BIP-CAT-TP configuration, while they may not have the "Applet for OTA pull" which requires a special order from the card manufacturer. This Applet in general is designed to poll periodically the OTA server; it is also possible to have it developed to provide a manual menu in the UE to search for SIM updates in the OTA server.

It must be stressed that the OTA-IP server is independent of the core network elements (GGSN-S/PGW) *and cannot know which IMSI corresponds to an IP source packet* when the "Applet for OTA pull" establishes the connection. Hence, any sequence of commands start by the interrogation of the *active IMSI* as illustrated in Figure 4.7. It could be the nominal IMSI or an auxiliary IMSI because resilience on a public MNO is active, and hence the OTA server *must know all the IMSIs of a given subscriber.*

All the above IP packets are ciphered and secured according to ETSI TS 102 225 [4.24] with the two keys Kic and Kid. At the end of the IP exchange HTTPs (TCP)or BIP/CAT-TP(UDP) according to ETSI TS 102 227 [4.21] (the same as in Figure 4.5 card initiated https), the SIM sends the PoR. In the example of Appendix 11.1, you see the full traces including the IP exchange.

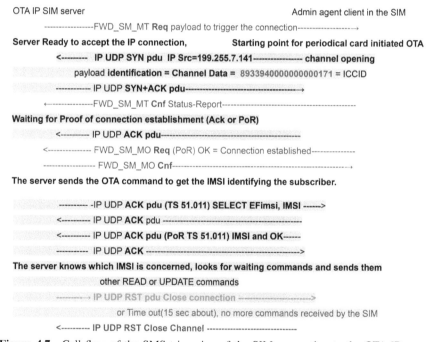

Figure 4.7 Call flow of the SMS triggering of the SIM connection to the OTA-IP server (UDP mode).

For a pure SIM initiated OTA IP, the messages in blue do not exist, but the IP sequence is the same.

The example shows that the OTA-IP can quickly perform a sequence of several READ and UPDATE in the same connection.

4.5.4 GSMa SP02 v3.2

This comprehensive specification has specified a certain number of secured control protocol

SCP80 which includes [4.1, Section 2.2.5] the methods presented above, that is:

- CAT-TP/UDP
- SMS push
- BIP/TCP which uses https.

4.6 Profile update of the security domain and protection against the cloning of a stolen SIM

When UICC SIMs are used of the type which allows to update the security domain, how to prevent any possibility of cloning the SIM, with a local tool (a PC with a SIM card reader) or using OTA if the admin password is known or the OTA 23.048 keys are known. The card administrator at the MNO could fraudulently try to obtain them.

> To *secure against the cloning* of a stolen IMSI or by an administrator, while letting the update possibility, it is sufficient to have the two files EF_KeyKi (which has Ki) and EF_Milenage_Parameters (which also has the OPc) *not readable* (only writeable) both using a local card administration tool, or using OTA.

The master OP key is not in the SIM card; it allows computing OPc from the Ki [4.13]. The IMSI can be read, but without the Opc, Ki, and the mileage parameters even if they are the standard values in most cases, the cloned IMSI will not be identified by the HLR-HSS. If one uses a non-standard milenage, it makes a *brute force attack impossible* with a total number of bits to be found of $128 + 128 + 2 \times 5 \times 128$ (Ki, Opc, 2×5 mileanage parameters).

The HLR-HSS *must also be protected from a fraudulent dumping* of the AuC database allowing us to have the Ki of a given IMSI. This is provided by an HLR-HSS which uses ciphered (e.g., AES) Ki with a key only known from the security management. In practice, the PMRs must be Full MVNOs with their own HLR-HSS as such a security is difficult to obtain from the HLR-HSS of a hosting MNO common to its own subscribers and to several MVNOs including the PMR.

4.7 Application provisioning in the device (not in the SIM card)

As seen in Chapter 2, it is useful to set the various groups with their own application and APN. This has nothing to do with the SIM card profiles. It can be done using the standard OTA GPRS methods [0.7, Chapter 9] and SMS-MT. A more comprehensive approach is the use of MDM systems in "pull mode" from the MDM suppliers such as PushManager.

The *application provisioning is a much more frequent operation than the rare changes in the SIM cards'* content.

4.8 Is being a full MVNO justified for an autonomous car manufacturer?

A car manufacturer in France or Germany, selling "connected" then "autonomous" cars in country C of Asia, has the choice between eUICC cards or UICC cards to select the local main MNO once the vehicle is in the final country. All data traffic will be paid by the subscriber of the IMSI which is the car manufacturer and charged as a service charge to the car owner.

An alternative discussed in this section is for the car manufacturer is to have its own IMSI and a PMR *type data only core network* with direct or "sponsored multi-imsi" roaming agreement in the various countries where its cars are sold, what would be the advantages?

4.8.1 Current high latency connected applications from the car to the manufacturer

Cars are increasingly connected through cellular services (see Figure 4.8). In Europe, any new car sold since April 2018 in the EU should now be compliant with E-call emergency services requiring an embedded cellular modem.

Drivers and passengers expect the presence of services already available on their smartphone, with a high quality, such as streaming video (live sport channels, on line gaming etc). They encompass generic services such as interpersonal communications: (including video conferencing), internet web portal and multimedia services. Beside the mentioned mandatory emergency services already stated, they also include car industry specific services such as:

- Secured driving services-help for driving
- Navigation-traffic/augmented reality/toll-parking/fleet management/geographical maps
- Preventive maintenance
- Experience feedeack/reliability
- Multimedia assistance in case of incident/accident (repair) or in case of emergency (help)
- Insurance services (tracking services, prevention of car theft); options such as insurance pay per km, pay how you drive.

These applications'response latency are *compatible with a roaming data transmission on the GRX network.*

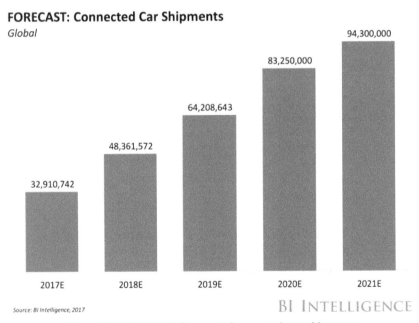

Figure 4.8   2017–2021 Forecast of connected cars shipments.

4.8.2 The next big thing: Autonomous vehicle with sensors

Soon autonomous vehicle will be available massively. All big brands will launch commercially autonomous cars within two years. For example, GM is expected to start mass production within a year. Here is a list of sensors in the Renault 2017 experimental autonomous vehicle.

1. Three LIDARs or long-range laser scanners (two front, one rear)
2. One long-range front RADAR
3. Four medium-range corner RADARs
4. Three digital cameras with short, medium and long focal lengths, at the top of the windscreen
5. Four 180° short-range digital cameras under the wing mirrors and at either side of the licence plate (parking assistance)
6. A belt of 20 SONARs (short-range ultrasound sensors) for parking assistance

> However the propagation delay on a GSM data network (in the GRX data network at least 50 msec in 2018, 5G will only speed up) excludes almost all distant real time control application. Collision control, headlamp control must be executed by a local computer in the car. Car parking (speed of 0.1 m/sec is the only one which could accept the 50msec) but a local computer does it well using the SONAR sensors.

In [4.26] it is stated by a main processor manufacturer that the data flow between the car and the processing will be 4000 Gbyte/day. But most of the trafic will be processed locally, not remotely by the manufacturer's application servers! For the debate of this section simply shows that the proceesing powe will tend to be what was availbale in an aircraft 20 years ago.

All those data may not need to be transmitted in real time to the cloud, but a substantial part would need: autonomous driving is based on accurate and very data hungry 3D maps that would require to be updated in near real time *with anticipation* in order to *circumvent the latency and the bandwidth practical limitation*, and car sensors would need to bring back their data in real time to the cloud to update such maps. *Being a full MVNO will not change the performance of the GRX data network.* Quoting[4.26]:

- In an autonomous car, we have to factor in cameras, radar, sonar(ultrasonic sensors for vehicle parking), GPS and LIDAR (Laser Imaging Detection and Ranging) – components as essential to this new way of driving as pistons, rings and engine blocks. Cameras will generate 20–60 MB/s, radar upwards of 10 kB/s, sonar 10–100 kB/s, GPS will run at 50 kB/s, and LIDAR will range between 10–70 MB/s. Run those numbers, and each autonomous vehicle will be generating approximately 4,000 GB – or 4 terabytes – of data a day.

- Every autonomous car will generate the data equivalent of almost 3,000 people. Extrapolate this further and think about how many cars are on the road. Let's estimate just 1 million autonomous cars worldwide – that means automated driving will be representative of the data of 3 billion people.

Current autonomous test vehicles (Volvo case) generate ony 4 Gbyte/hour which are processed in the local computer.

4.8.3 Data trafic costs comparison between Local IMSI and full MVNO

The rationale to build its own MVNO with its own IMSI (or a sponsored IMSI) in the car modem *cannot be economic in general.* The trafic is paid by a service subscription to the car manufacturer and cannot exceed a few monthly USD, which is less than the price of 10 Mbytes of roaming data for certain networks. The car manufacturer would be quickly broke.

An economic model relying on the data roaming can only work in certain regulated situations such as Europe where the APN internet.eu is uniformely charged with no difference between HPLMN and VPLMN tariffs. Even in this case, a local IMSI may provide important price savings though the negociation of M2M special rates. We can conclude that the car manufactuter's current strategy of setting local IMSIs instead of being a full MVNO is optimal from a cost prospective.

4.8.4 Security discussion: Local IMSI compared to own IMSI as a full MVNO

The car manufacturer of connected or later autonomous vehicles wants to be protected against the hacking of the car: thief, malicious blocking of the braking system, etc. Attacks using the bluetooth maintenance interface have been reported since 2016, here we consider attacks coming through the data interface of the modem from a man in the middle. The use of a secured ciphered IP protocol between the device in the car and the car manufacturer's server guarantees the security aganst interception.

The security issue concerns then the protection against re-routing of the data flow to a hacking site which would insert a spyware in the car device to take its control using a weakness of the software as exists in many browser software. The car devices have an IMEI which is characteristic of the M2M type devices, this means that the traffic of a specific IMEI car be routed to the spyware installation site with either an IMSI catcher, or the Legal Interception system (see Chapter 8). The stratregies for country A which wants to be in a position to control all the cars could be:

- all the cars sold in the country have the IMEI and IMSI declared to the authorities, this is the simplest,
- an IMSI catcher with the MCC-MNC of all teh MNOs is installed close to all the car vending offices.

Wether the cards use an IMSI of one the MNOs of A or the IMSI of the car manufacturer full MNO's it is the same to obtain the IMEI and IMSI. Only visiting cars would escape the scheme provided that at the border there is no IMSI catcher.The only general protection that the car manufacturer can provide is to use only secured OS for each SW releases.

4.8.5 Supplementary features provided by the full MVNO model

In any case Full MVNO means that there is a single HLR-HSS for all the countries and including various IMSIs used. There are 3 ways to consider "various IMSIs", each *use a central HLR-HSS for all the countries*:

1. "multi-IMSI solution": The SIMs has various "local IMSIs" *allocated by a local MNO* which are remotely selectable (domain switch), then handled by the central HLR-HSS.
2. *"local full MVNO"* roaming: local IMSIs and Full MVNOs agreeements in each country are negociated. Example Lycamobile which is an "IMSI integrator" in the Netherlands negociates IMSIs *from the local regulator* in most european countries with *local agreements with one operator ar least*. It is said (2018) that such a setup can be done in about 2 months, remembering that no equipment needs ever to be installed in the visited country. The central HLR-HSS needs only SS7 and diameter connections with the visited networks.
3. in restricted countries one is obliged to use a local HLR-HSS with a local IMSI, this forbids the full MVNO mo with an HLR-HSS in another country.

The full MVNO can provide the car manufacturer with supplementary features compared to traditional solutions:

- An independent MVNO being MNO agnostic could hook to several MNO in a single country. Google MVNO Project Fi in the US is based on this principle, whereby it switches automatically to hook to the best signal between T-Mobile, Sprint and US Cellular. If emergency services (E-call) have access to all cellular accesses, this is not the case for all other services. Always Best Connected will be key for example for assistance services or map downloading. UK and the Netherlands have already allowed formally those multi-hosting services. Compared to other existing device based services, the full MVNO can proactively and dynamically switch to the best available network at any time without disruption.
- Car manufacturers are dependent on the service design made by their traditional MNO partner to build their own services: for example it is clear that the edge computing close to the vehicle will be necessary for the autonomous car: the solutions and implementation speed from MNOs could vary and may provide unsatisfactory results, a specific solution by car manufacturers requiring a network would guarantee a suitable service.

4.8.6 Minimum setup for a car manufacturer to manage their SIMs: OTA-IP server

4.8.6.1 Need to have its own OTA server for its own management of the SIMs and the SW updates

Let us first consider that the car manufacturer should *manage directly the setup of their SIMs* with various IMSIs from the visited operators, meaning they must have their own OTA server and not depend on the SIM manufacturer's which security must always be questioned (this is the case for all examples of Table 4.3, including the case where they have a direct interface with the OTA server of the "IMSI aggregator"). They need:

- to be the "editor" of the SIM that is being able to specify in detail the "electric profile", with all the directories, files and contents, and the applets. They should have specialised engineers to do it.
- they should buy an OTA server using the OTA-IP as it does not require to have a SMSC and a SS7 and/or GRX subscription, only a secured

internet connection. The *cards should have the "polling applet"* of Section 4.5.1.2 which is in charge to establish the BIP/CAT-TP channel (every few hours) or the http(s) connection with the OTA-IP server of the card manufacturer.

- they do not need to have their own IMSI; in the design of the SIM, two security domains only must be defined: one for a "nominal IMSI" of the manufacturing country with many roaming agreements (e.g. Orange France, Deutsch Telecom, etc..), the second for the "auxiliary" IMSI set remotely with OTA-IP of this chapter, when the car reaches the country where it is used. This means that applying for a MNO or full MVNO license to the local regulator for the "nominal IMSI", is not required. They just need to have commercial *Full MVNO agreements* with the various MNOs concerned to have an allocation of IMSIs which would be managed by the car manufacturer's system including the HLR-HSS.

Having its own even expensive OTA system for the update of SW (not the SIM parameters) is mandatory to avoid the call back of cars which is a very heavy operation.

Table 4.3 Summary of 2018 M2M solutions of car manufacturers negociated or contracted, deployment just starting

Car Manufacturer	IMSI Agregator (Solution 2) with the HLR-HSS	SIM Manufacturer	SIM Editor (who Orders and has the OTA)	Multi-Imsi or Local Full MVNO	Europe Only	Resticted Countries (Solution 3) Russia, China, Brazil
Volkswagen Audi	CUBIC	Gieseke& Devrient	CUBIC	Multi-IMSI	Yes	HLR-HSS and IMSI of the country
Fiat	Transatel	Imedia	Transatel	Multi-IMSI	Yes	HLR-HSS and IMSI of the country
Jaguar Land Rover	Transatel	Imedia	Transatel	Multi-IMSI	Yes	HLR-HSS and IMSI of the country
GM	Globe Touch(India)	Not known	Not known	Multi-IMSI	Outside US	HLR-HSS and IMSI of the country
Kia	True Phone(UK)	Not known	Not known	Multi-IMSI		HLR-HSS and IMSI of the country

4.8.6.2 Consequence: The card manufacturer must be a full MVNO

The card manufacturer having its own HLR-HSS is a consequence of being able to setup the SIMs, *as using the HLR-HSS of the MNOs would not provide the Kic, Kid, Kik OTA security keys* necessary to update the SIMs in particular adding an auxiliary IMSI's security domain.

There are 3 main possibilities including a mixed case:

- the card manufacturer: 1) develops its own Full MVNO, buys a proven core network (1rst tier equipement providers such as Nokia, Ericsson, Cisco, the chinese manufacturers if they are allowed to provide the core network) including a HLR-HSS as a minimum in a managed mode (the 1rst tiers may be quite more expensive than HLR-HSS specialised manufacturers but their long term existence will not be challenged), 2) *negociates directly* commercial agreements with the foreign MNOs. This is a very long and specialised work with a roaming agreement team needed.

- 2) the car manufacturer relies on an "IMSI agregator" which provides its own core network. This approach has been selected by some car manufacturers such as Audi, Volkswagen, GM, KIA, Fiat, Jaguar, Land Rover, but may be considered as a provisional choice as these aggregates do not have the development backing to guarantee the evolution to 5 and their long term existence currently supported by Venture capitalists with dumped prices to create markets. These core network may not be considered as "car manufacturer grade" in terms of robustness and evolutionary.

- a mixed case which is probably a good compromise and allows to migrate to "full MVNO" while allowing to start quickly and cheaply is to rely on the "SIMS aggregator" to have the auxiliary IMSIs ranges and their agreements. But the car manufacturer would have its own HLR-HSS(3G-4G) then AUSF-UDM(5G), would create the SIM cards and would be the only one to know the above OTA security Keys (Ki, Opc, Kic, Kid, Kik would not be known by the agregator, only the IMSI ranges). The HLR-HSS and OTA system are the only core equipments needed by the car manufacturer; the GGSN-PGW, and billing would be the agregators.

4.8.6.3 Summary table of the 2018 solutions for car manufacturers

As shown by the above table, the "local full MVNO" which many favor for car manufacturers does not seem yet ready to be deployed although the negociation of a local full MVNO licence with the regulator and a national network usage agreement is in practice easier that negociating the obtention of a range of IMSIs from local MNOs.

In certain countries (China, Russia, Brazil) the use of a HLR-HSS ouside the country is forbidden. There is no other way than to have a local subscription handled by the local core network of the operator. Solution 1) does not seem to be yet used.

References

[4.1] Gsma, "Remote Provisioning Architecture for Embedded UICC Technical Specification," version 3.2, 27 June 2017, *309 pages.*

[4.2] 3GPP TS 31.102 v14.4.0 (2018-07), "Universal Mobile Telecommunications System (UMTS); LTE; Characteristics of the Universal Subscriber Identity Module (USIM) application," Release 14, *the full list and characteristics of all directory and files in a SIM card. Includes the recent files for MBMS, MCPTT, ProSE, Multimedia, etc.*

[4.3] 3GPP TS 31.101 V14.1.0 (2017-04), "Universal Mobile Telecommunications System (UMTS); LTE; UICC-terminal interface; Physical and logical characteristics," Release 14.

[4.4] 3GPP TS 23.122 v14.4.0 (2017-10), "Digital cellular telecommunications system (Phase 2+) (GSM); Universal Mobile Telecommunications System (UMTS); LTE; Non-Access-Stratum (NAS) functions related to Mobile Station (MS) in idle mode," Release 14. *This is the most important standard for resilience in PMR networks.*

[4.5] 3GPP TS 23.048 v5.9.0 (2005-06), "Digital cellular telecommunications system (Phase 2+); Universal Mobile Telecommunications System (UMTS); Security mechanisms for the (U)SIM application toolkit; Stage 2," Release 5. *Gives a (very) summary of the Applet Management Commands (INSTALL, LOAD, GET DATA, etc.), [4.8 and 4.9] are the full references. See page 17, very close to TS 102 225[4.24] for the secured packet (OTA-IP), the CPI for the commands is '70' instead of '01', the RPI is '71' instead of '02'. Only 3DESCBC, no AES available otherwise.*

[4.6] 3GPP TS 51.011 V4.15.0 (2005-06), "Digital cellular telecommunications system (Phase 2+); Specification of the Subscriber Identity Module – Mobile Equipment (SIM-ME) interface," Release 4, *All the OTA commands to SELECT, READ, UPDATE BINARY, UPDATE RECORD, GET RESPONSE, etc. the various SIM files in TS 31.102. Replaces the old TS 11.11 V5.3.0 (1996-07).*

[4.7] 3GPP TS 31.111 V14.4.0 (2017-10), "Digital cellular telecommunications system (Phase 2+) (GSM); Universal Mobile Telecommunications System (UMTS); LTE; Universal Subscriber Identity Module (USIM) Application Toolkit (USAT)," Release 14, *replaces the old TS 11.14. Explains the "proactive commands" (DISPLAY TEXT, REFRESH, RUN GSM, SETUP CALL, SEND DTMF, ENVELOPE, etc.) which are used in a SIM Tool Kit to interact between the device and the SIM card.*

[4.8] W. Rackl, W. Effling, "Smart Card Handbook," Wiley, 2010. *Two authors from Giesecke and Devrient, show to LOAD.cap files as packages (applet). A good reference book on Applet Management.*

[4.9] Open Platform Card Specification version 2.0.1, available at: http://www.globalplatform.org

[4.10] ETSI TS 102 226 V13.0.0 (2016-05), "Smart Cards; Remote APDU structure for UICC based applications," (Release 13), *describes the commands for Remote Applet Management (RAM), see 4.3.1.3; see section 9.2 for the coding of the channel activation SMS Push commands; see the command structure RFM over HTTP(s).*

[4.11] ETSI TS 102 223 V14.0.0 (2017-05), "Smart Cards; Card Application Toolkit (CAT)" (Release 14), *Smartcard equivalent of [4.7] of "proactive commands" previously TS 11.14 plus OPEN CHANNEL, CLOSE CHANNEL, etc.*

[4.12] ETSI TS 102 221 V13.1.0 (2016-05), "UICC-Terminal interface; Physical and logical characteristics UICC-Terminal interface; Physical and logical characteristics," (Release 13), *contains the MANAGE CHANNEL commands.*

[4.13] 3GPP TS 35.206 V13.0.0 (2016-01), "Universal Mobile Telecommunications System (UMTS); LTE; 3G Security; Specification of the MILENAGE algorithm set: An example algorithm set for the 3GPP authentication and key generation functions f1, f1*, f2, f3, f4, f5 and f5*; Document 2: Algorithm specification." *Description of the algorithms and personalisation of the 2×5 parameters ci et ri milenage.*

[4.14] RFC 4279, "Pre-Shared Key Ciphersuites for Transport Layer Security (TLS)," December 2005, *Section 2, page 4.*

[4.15] TS 23.057 v14.0.0 (2017-05), "Digital cellular telecommunications system (Phase 2+) (GSM); Universal Mobile Telecommunications System (UMTS); Mobile Execution Environment (MExE); Functional description; Stage 2," Release 14.

[4.16] RFC 5246, "The Transport Layer Security (TLS) Protocol Version 1.2," January 2008.

[4.17] ETSI TS 101 220 V12.0.0 (2013-10), "Smart Cards; ETSI numbering system for telecommunication application providers" (Release 12), gives the TAGS to encode commands of TS 102.226.

[4.18] "Recommandation de sécurité relatives à TLS," ANSI 2016, pages 45–52.

[4.19] TS 36.306 v14.5.0 (2018-01), "LTE; Evolved *Universal* Terrestrial Radio Access (E-UTRA); User Equipment (UE) radio access capabilities," Release 14.

[4.20] 3GPP TS 23.040 V13.1.0 (2016-04), "Digital cellular telecommunications system (Phase 2+) (GSM); Universal Mobile Telecommunications System (UMTS); Technical realization of the Short Message Service (SMS)," Release 13.

[4.21] ETSI TS 102 127 V6.13.0 (2009-04), "Transport Protocol for CAT Application," Release 6. *End-to-end UDP based protocol between a SIM card and an OTA server.*
OTA-IP security

[4.22] ETSI TS 131 115 V14.0.0 (2017-04), "Digital cellular telecommunications system (Phase 2+) (GSM); Universal Mobile Telecommunications System (UMTS); LTE; Secured packet structure for (Universal) Subscriber Identity Module (U)SIM Toolkit applications," Release 14. *Appeared in 2005.*

[4.23] ETSI TS 131 116 V14.0.0 (2017-04), "Digital cellular telecommunications system (Phase 2+) (GSM); Universal Mobile Telecommunications System (UMTS); LTE; Remote APDU Structure for (U)SIM Toolkit applications," Release 14. *Appeared in 2008.*

[4.24] ETSI TS 102 225 V12.1.0 (2014-10), "Smart Cards; Secured packet structure for UICC based applications," Release 12. *See page 17, very close (Packet OTA case) to TS 23.048 (SMS-PP case)f, the CPI (Command Packet identifier) is just '01' instead of '70', the RPI is '02' instead of '71'. Covers BIP/CAT-TP and HHTPs. AES introduced for the Kic parameter.*

[4.25] ETSI TS 101 220 V15.0.0 (2018-01), "Smart Cards; ETSI numbering system for telecommunication application providers," Release 15, *coding of the lengths of secured packets in [4.24] or [4.5] and definition of the tags used in section 13.1 for the coding of 102 223 commands.*

[4.26] https://www.networkworld.com/article/3147892/internet/one-autono mous-car-will-use-4000-gb-of-dataday.html, *opinion of the Intel CEO on the data volumes generated by the autonomous vehicle.*

5

Group Communication Provisioning by OTA, SMS 4G, and SMS IMS

Horum omnium fortissimi sunt Belgae, propterea quod a cultu atque human-itate provinciae longissime absunt, minimeque ad eos mercatores saepe commeant atque ea quae ad effeminandos animos pertinent important.

Of all these (peoples), the Belgians are the bravest, because they are furthest from the civilization and politeness of our provinces, and merchants least frequently visit them, and import those things which tend to effeminate the courage.

Commentarii de bello gallico, Julius Cesar, J.M.Eberhart, editor, Paris, 1809, Volume 4, Liber 1, Chapter 1, page 4.

5.1 Operational need for OTA provisioning in PMR networks

This chapter's quote is a tribute to the first operational European LTE-based PMR by ASTRID the Belgium federal telecommunication agency in 2013; adding LTE with access priority to the public MNOs in 2016, they had a full OTA system to perform the changes when needed in the SIM cards and terminals.

Chapter 2 has explained that the radio coverage of the security forces may be the result of the federation of several tactical networks and of the public MNOs' coverage. The current organization with groups and their leader is kept, while the LTE equipment themselves are not assigned to a particular group; they select one as available. The security forces may be heavily equipped for their protection with heavy gloves and are not selected for their computer usage skills. Hence, the handsets of a group must all be provisioned at the beginning of a mission with the ID of the group corresponding to the ID of their tactical LTE network if they use one. The most standard way to do this is to have an APN for each group, the same as the MCPTT application of that group.

Speed and simplicity lead to perform the provisioning automatically with an OTA (for GPRS profile) server which is then part of each tactical network.

OTA uses SMS-MT, and hence there must be an SMS capability in the 4G tactical networks, which means prerequisites in the architecture of the tactical network.

- an SMSC with an SM_IP_GW (delivery in 4G);
- the HLR-HSS which must have the S6c/Diameter for SEND_ROUTING_INFO_FOR_SM;
- the MME which must have the SGd/Diameter interface for SMS.

5.2 SMS service convergence 2G, 3G, 4G, SIP, and SMPP in other Non-PMR cases

There is more generally the need to allow SMS (the SIP equivalent uses SIP MESSAGE) from/to various sources or destinations including mobiles visiting 3G networks, pure 4G networks, "or IMS networks" using SIP. A comprehensive 3GPP architecture is described in this chapter which *also* provides *MMS interworking* between SIP and 3G beyond the 3GPP IP-SM-GW function [3.6].

In this chapter, we provide a unified interworking view which includes the IP-SM-GW function to interwork SIP messages with 3G and 4G messages. We explain with details the least known of SMS: SMS over 4G (without CFB) using an IWF-SMS and SIP ↔ MAP SMS interworking. The IP-SM-GW function has appeared in 2010 in the MAP standard TS 29.002 to allow the SIP ↔ MAP SMS convergence.

5.3 SMS in the EUTRAN 4G domain

In 2017, there was a confusion on the SMS sending/receiving procedure when a user is registered in 4G. Some still believe that the CSFB is used, the SMS-MO is sent in 3G, and the SMS-MT would be received in 3G too.

This is wrong; the UE *remains active in 4G (the UE does not switch to 3G)* for SMS-MO and SMS-MT reception. There are two possibilities:

- a CSFB, for example, SMS-MT; it is received in 3G by the MSC/VLR associated to the MME which relays it to the MME with the SGs-AP/Diameter interface and then 4G SMS starting with paging is used;

- the MME receives the SMS-MT on the SGd/Diameter interface from the Diameter capable SMSC.

Similarly, for the SMS-MO:

- If the MME does not have the SMS capability (SGd/Diameter), it relays the SMS-MO to the MSC/VLR.
- If it does, it sends it directly to the Diameter capable SMSC.

In any of the two cases, the UE *does not perform a CSFB to 3G*.

There is also another frequent misunderstanding that because if the MME is SMS capable, it sends a GT in the UPDATE_LOCATION that would mean that there is MAP SS7 in the MME. *This is wrong* and the purpose is to indicate only the SMS capability as explained in the next section.

In Section 5.6, the traces show that the segmentation is possible in 4G identical to 3G, the 3G SMS MAP format being encapsulated in a Diameter transport envelop which makes possible the implementation of the IWF-SMS function in Figure 5.1.

The case of SMS direct capability with SGd/Diameter in the MME is more efficient (faster delivery) and more secured for the assurance of delivery or not delivery, and applies to the many VoLTE networks which do not have CSFB and are pure 4G, as well as to all the pure 4G PMR networks. For SMS hub operators, it also avoids sharing the revenues with "peers" the real quality

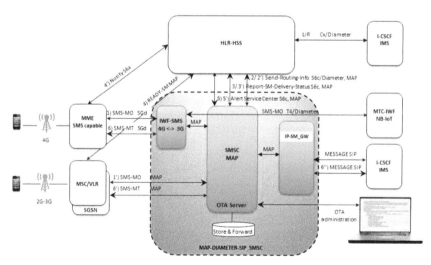

Figure 5.1 4G Diameter capable SMSC with an IWF-SMS (SMS 4G ↔ 3G), IoT interface, and SIP interface IP-SM-GW.

of which is unknown even when they respond with an end delivery being performed. For a major hub operator, it is an aim to have direct delivery. It classical and well known with GSM MAP SS7 (covered in [0.7]), less known, know largely obsolete with IS-41 (covered in [0.9, Chapter 10]). These are legacy protocols which will be superseded completely by the Diameter protocols in less than 5 years as the equipment become obsolete and are replaced for most destinations' SMS. Hence, this topic will soon become major. This chapter gives a reference technical presentation of the subject.

5.4 SMS procedure to handle destinations in 4G networks

5.4.1 SMS procedure and call flow

An SMSC-capable SMS may be implemented with a classical MAP SS7 SMSC and a separate IP-SM-GW with Diameter or a fully integrated system. It can interrogate the HLR-HSS with MAP or Diameter Send-Routing-Info-For-SM.

Figure 5.1 gives an example of implementation using the case of a handset visiting the MME of a 4G network and sending an SMS to a handset using 3G visiting an MSC/VLR or an SGSN with SMS enabled:

- Sending a 1) 4G SMS to a 3G customer (he does not know). It goes to the SMSC through the IWF-SMS which converts to the internal MAP SMS format in this implementation example.
- The SMS interrogates the HLR-HSS of the destination using 2) Send-Routing-Info for SM Diameter or 2') MAP. The destination is unreachable:
- The SMSC sends a Report_SM_Delivery_Status 3) Diameter or 3') MAP to the HLR-HSS in order to be alerted when the destination becomes reachable.
- When this is the case, the MSC/VLR or the SGSN sends a 4) READY-FOR-SM MAP to the HLR-HSS. If the unreachable destination was 4G, the MME would send a Notify S6a/Diameter 4').
- Immediately, the HLS-HSS sends 5') Alert-Service-Center S6c or 5) MAP to the SMSC.

The SMSC retrieves the message from its Store&Forward, interrogates the HLR-HSS with a SRI_FOR_SM Diameter or MAP, finds in the response (see Section 5.4.3) that it is not visiting a 4G MME, and then sends

6) MT-Forward-SM directly in MAP. If it was visiting an MME (the MME_diam_name is set), it would send 6') through the IWF-SMS to convert to SGd/Diameter. If it was SIP registered (the MAPPN_ims_node_ind is set), it would send the message 6''') in SIP with a MAP→ SIP conversion performed by the IP-SM-GW.

The "universal SMSC" *has all the standard SMS submission interfaces*

- MAP and Diameter with 3G and 4G MNOs
- SMPP with aggregators, peer hubs, or MNOs using SMPP (few)
- T4 with NB-IoT operators (see Chapter 10) from their SCEF.

It is able to interface both ways with SIP MESSAGEs. When a concatenated SMS-MT is received from 3G or 4G, it will buffer the segments until all are received and then send a single long SIP MESSAGE to the SIP registered user. It will perform the necessary translations from GSM 7-bit alphabet [5.7] to ISO 8859-1 then to UTF8 compatible with any SIP client and including all the special characters.

Not shown is the optional Sh interface to interrogate the HLR-HSS with a UDR/Sh/Diameter (User-Data-Request) which allows getting the detailed user profile for the SMS service.

In order to provide the same service as in 3G (store&forward when the destination is not reachable) when an SMS (SIP MESSAGE) is sent from SIP to SIP, the SIP MESSAGE *is not sent directly* and the IP-SM-GW *is used as a proxy*, as it is connected to the SMSC which provides the *store and retry function* as part of the overall SMSC whether *the destination is SIP, 3G, or 4G registered.*

The OTA application is connected to the OTA server. It can use the SMSC but also use IP with BIP/CATP or HTTPS protocols as explained in Chapter 4.

5.4.2 Virtualized type 1 implementation example

Two physical redundant servers only will be enough to provide all the different functions. As can be seen, according to the rules in [0.10, Chapter 11], IWF-SMS, SMSC, and IP-SM-GW *which all three receive MAP* cannot be on the same VM as MAP SMS format is used to unify between SMS 4G, SMS 3G, and MESSAGE SIP.

The call flow explanation is given by Figure 5.1 showing that there is a common Store&Forward of the SMS whether there are deposited by SMS 4G, 3G, or MESSAGE SIP.

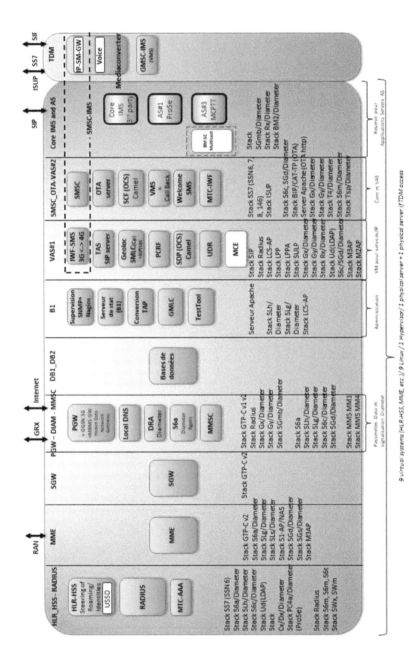

Figure 5.2   Implementation of the Diameter-capable SMSC with a virtual machine architecture.

5.4.3 HLR-HSS interrogation with MAP/SS7 (3GPP TS 29.002)

If the SMSC uses MAP, it sends (2') a SEND_ROUTING_INFO_FOR_SM Req to the HLR-HSS. The support of MME delivery has appeared in MAP Rel 11 of [3.1].

```
- - - - Super Detailed SS7 Analyser (C)HALYS (Trace Level 8) - - - -
   PA_Len = 28
   MAP-SEND-ROUTING-INFO-FOR-SM-IND(2)
     MAPPN_invoke_id(14)
       L = 001
       Data: 1
     MAPPN_msisdn(15)
       L = 007
       Data: Ext = No extension
             Ton = International
             Npi = ISDN
             Address = 33970675065
     MAPPN_sm_rp_pri(16)
       L = 001
       Data: (1):High Priority
     MAPPN_sc_addr(17)
       L = 007
       Data: Ext = No extension
             Ton = International
             Npi = ISDN
             Address = 33970670001
     MAPPN_GPRS_Support_Indicator(118)
       L = 000
     MAPPN_IP_SM_GW_GuidanceIndicator
       L = 000

- - - - Super Detailed SS7 Analyser (C)HALYS (Trace Level 8) - - - -
   PA_Len = 151
   MAP-SEND-ROUTING-INFO-FOR-SM-RSP(129)
     MAPPN_invoke_id(14)
       L = 001
       Data: 1
     MAPPN_imsi(18)
       L = 008
       Data: Address = 208940000001065
     MAPPN_msc_num(19)
       L = 007
       Data: Ext = No extension
             Ton = International
             Npi = ISDN
             Address = 33970670007
   ip_sm_gw_num(GT of SMS capable MME)(832)   // if the UE is
                                                 SIP registered
       L = 007
       Data: Ext = No extension
             Ton = International
             Npi = ISDN
```

```
            Address = 33970671234
MAPPN_ims_node_ind(826)                          // if the UE is
             SIP registered
   L = 000
MME_diam_name(820) // 'MME_name'                  // if the UE is 4G
                                                     registered
   L = 040
   Data: ipsmgw1.epc.mnc094.mcc208.3gppnetwork.org  // from the
                                                     ATM Ind below
   L = 034
   Data: epc.mnc094.mcc208.3gppnetwork.org          // from the
                                                     ATM Ind below

MAPPN_ellipsis(57)
   L = 015
   Data: (Hex) 010D696D732E68616C79732E667200
         HALYS_ellipsis(0x01) IMS server name = ims.halys.fr
```

In the SEND_ROUTING_INFO_FOR_SM Cnf, the GT (ip_sm_gw_num) of the MME is given as well as Diameter RFC 3588 Name and Realm from the Update Location Diameter. When it is included, it tells that SMS 4G delivery is supported. To get the MME delivery information, the SMSC has indicated a 4G capability with "IP_SM_GW_Guidance_Indication."

The response below is illustrative but possible (a destination UE which is 3G, 4G, and SIP registered).

From the response SRI_FOR_SM Cnf, the SMSC sees that the UE is registered both in 3G (for CSFB), also on the MME mme01.epc.mnc094.mcc208.3gppnetwork.org and also on the SIP ip_sm_gw. It may decide which delivery mode it prefers.

5.4.4 HLR-HSS interrogation with S6c/diameter (3GPP TS 29.338)

The protocol is Diameter with AVPs but the semantic content of the request and of the response is exactly the same. Obviously, it will replace the MAP interrogation in the close future. This corresponds to 2) in Figure 5.2 where Diameter is used instead of MAP 2').

5.4.5 SIP registration in the SM-IP-GW to receive SMS with SIP MESSAGES

5.4.5.1 Standard 3GP registration for SIP message reception

Annex B3 [5.9] describes the standard 3GPP way used to register SIP UE to be able to receive SMS-MT. When a REGISTER is received by the P-SCSF, it sends SUBSCRIBE to the SM-IP-GW which will register on the HLR-HSS using a MAP ANY-TIME-MODIFICATION Req (ARM) containing the gsmSCF_address (the GT of the IP-SM-GW). The role of the ATM to register the reachability of a UE for SIP SMS has appeared in Rel 7 (2010-06) of [5.1] with the IMSI or MSISN identities also replaced by the more general subscriber identity.

1. Successful UE registration
2. Evaluation of initial filter criteria   // reads the profile using UDR/Sh
3. REGISTER
4. 200 OK
5. SUBSCRIBE
6. 200 OK
7. NOTIFY $<$state = "**active**"$>$ I-CSCF $\rightarrow$ IP-SM-GW
8. 200 OK
9. MAP:ATM   // IP-SM-GW $\rightarrow$ HLR-HSS writes the IP-SM-GW,
            parameters are:
                                // Subscriber Identity (MSISDN),
                                // modification request for IP-SM-GW
                                    data,
                                        // gsmSCF-address(E164 of
                                            IP-SM-GW)
                                // **activate** request for UE reachability
                                    = reachable
10. MAP:ATM response HLR-HSS $\rightarrow$ IP-SM-GW
The detailed MAP trace is in the next section.

5.4.5.2 MAP traces for ANY-TIME-MODIFICATION IP-SM-GW $\rightarrow$ HLR-HSS

```
- - - - Super Detailed SS7 Analyser (C)HALYS (Trace Level 8) - - - -
        PA_Len = 65
        MAP-OPEN-IND(2)
          dest_address(Q713)(1)
            L = 015
            Data:   Route on GT, Global Title included(0x04),
                    Signalling Point Code (ITU) = 0-097-3 (  779)
                    Subsystem Number = HLR(6),
                    Global Title :
                        Translation Type = 0,
                        Numbering Plan = ISDN/Telephony(E164),
                        Encoding Scheme = BCD, odd number of digits,
                        Nature Address Indicator = International number,
                        Address information = 339706700000010
```

```
          orig_address(Q713) (3)
            L = 013
            Data:  Route on GT, Global Title included(0x04),
                   Signalling Point Code (ITU) = 1-201-1 ( 3657)
                   Subsystem Number = MSC(8),
                   Global Title :
                       Translation Type = 0,
                       Numbering Plan = ISDN/Telephony(E164),
                       Encoding Scheme = BCD, odd number of digits,
                       Nature Address Indicator = International number,
                       Address information = 33970670007
          QOS(238)
            L = 001
            Data: (Hex) 02
          MAPorCAP_Application_Context (11)
            L = 009
            Data: (Hex) 060704000001002B03
                    MAP_Any Time Info Handling Package_v3 MAP V3
          MAPPN_ellipsis(57)
            L = 015
            Data: (Hex) 820A53434F696C9059689B2E830103
                    HALYS_ellipsis(0x84) Origin PC = 0

 - - - - Super Detailed SS7 Analyser (C)HALYS (Trace Level 8) - - - -
       PA_Len = 113
       MAP-ANYTIME-MODIFICATION-IND (116)
         MAPPN_invoke_id (14)
           L = 001
           Data: 1
         MAPPN_imsi (18)
           L = 008
           Data: Address = 208940000001065
         MAPPN_gsmscf_addr (51)
           L = 007
           Data: Ext = No extension
                 Ton = International
                 Npi = ISDN
                 Address = 33970670001
         MAPPN_Modify_Registration_Status (521)
           L = 001
           Data: (1):Activate
         MAPPN_Modify_Request_IP_SM_GW ellipsis (540)
           L = 078
            HALYS_ellipsis(0x01)    IP-SM-GW     Diameter     name
ipsmgw1.epc.mnc094.mcc208.3gppnetwork.org
            HALYS_ellipsis(0x02)    IP-SM-GW     Diameter     realm
epc.mnc094.mcc208.3gppnetwork.org

   - - - - - - - - - - - - - - - - -- - - - - - - - - - - - - -

    ---------------------- HDR ------------------------ [ 20170813
_151241_954628 (954627 us) ]
# ROUTER # <=  MTU_ANY_TIME_MODIFICATION_IND:
```

```
                451 36149 +33970670007|0| 8 +0 +3397067000000010|0| 6 +0
0 3 3657 779 +33970670001 +0 +208940000001065 1 ipsmgw1.epc.mnc094.
mcc208.3gppnetwork.org
epc.mnc094.mcc208.3gppnetwork.org 00


    ---------------------- HDR ------------------------- [ 20170813
    _151241_954667 (954666 us) ]

    MAPE-R Instance = 0
    SS7E-R Type = MAP_MSG_DLG_IND (000087E3)
    SS7E-R Dialog_ID = 36149
    SS7E-R Src  = 15
    SS7E-R Dst  = 1D
    SS7E-R Rsp_req = 0  Status = 0  Err_info = 0
    - - - - Super Detailed SS7 Analyser (C)HALYS (Trace Level 8) - - -
        PA_Len = 2
        MAP-DELIMITER-IND (6)
# ROUTER # =>  MTU_ANY_TIME_MODIFICATION_RSP
            453 36149 0 0 0 0 +0 +0 00
RCV MESSAGE : 453 36149 0 0 0 0 +0 +0 00    ------------- HDR -------
------------------- [ 20170813_151241_958460 (958459 us) ]
    MAPE-E Instance = 0
    SS7E-E Type = MAP_MSG_DLG_REQ (0000C7E2)
    SS7E-E Dialog_ID = 36149
    SS7E-E Src  = 1D
    SS7E-E Dst  = 15
    SS7E-E Rsp_req = 0  Status = 0  Err_info = 0
    - - - - Super Detailed SS7 Analyser (C)HALYS (Trace Level 8) - - -
        PA_Len = 16
        MAP-OPEN-RSP (129)
          MAPPN_result (5)
            L = 001
            Data: (0):Accept
          MAPorCAP_Application_Context (11)
            L = 009
            Data: (Hex) 060704000001002B03
                  MAP_Any Time Info Handling Package_v3 MAP V3
    - - - - - - - - - - - - - - - -- - - - - - - - - - - - -
    ---------------------- HDR ------------------------- [ 20170813
    _151241_958595 (958595 us) ]
    MAPE-E Instance = 0
    SS7E-E Type = MAP_MSG_SRV_REQ (0000C7E0)
    SS7E-E Dialog_ID = 36149
    SS7E-E Src  = 1D
    SS7E-E Dst  = 15
    SS7E-E Rsp_req = 0  Status = 0  Err_info = 0
    - - - - Super Detailed SS7 Analyser (C)HALYS (Trace Level 8) - - -
        PA_Len = 5
        MAP-ANYTIME-MODIFICATION-RSP (231)
          MAPPN_invoke_id (14)
            L = 001
            Data: 1
    - - - - - - - - - - - - - - - -- - - - - - - - - - - - -
```

```
--------------------- HEX ------------------------ [ 20170813
_151241_958678 ]
--------------------- HDR ------------------------ [ 20170813
_151241_958701 (958700 us) ]
    MAPE-E Instance = 0
    SS7E-E Type = MAP_MSG_DLG_REQ (0000C7E2)
    SS7E-E Dialog_ID = 36149
    SS7E-E Src  = 1D
    SS7E-E Dst  = 15
    SS7E-E Rsp_req = 0  Status = 0  Err_info = 0
    - - - - Super Detailed SS7 Analyser (C)HALYS (Trace Level 8) - - -
        PA_Len = 5
        MAP-CLOSE-REQ (3)
          MAPPN_release_method (7)
```

5.4.5.3 Standard 3GP deregistration for SIP message reception

Annex B4 [5.9] describes the standard deregistration procedure using MAP ATM also.

1. UE deregistration (PURGE UE, etc.) → HLR-HSS
2. NOTIFY   S-CSCF → IP-SM-GW <state = "**terminated**">
3. 200 OK   IP-SM-GW → S-CSCF
4. MAP:ATM   // Subscriber Identity (MSISDN),
                          // modification request for IP-SM-GW
                             data,
                               // gsmSCF-address(E164 of
                                  IP-SM-GW)
                            // **deactivate** request for UE
                             reachability = not reachable

10. MAP:ATM response

The MAP trace for this ATM as received by the HLR-HSS is:
- - - - Super Detailed SS7 Analyser - - - - - - -

```
        PA_Len = 113
        MAP-ANYTIME-MODIFICATION-IND (116)
          MAPPN_invoke_id (14)
            L = 001
            Data: 1
          MAPPN_imsi (18)
           L = 008
            Data: Address = 208940000001065
          MAPPN_gsmscf_addr (51)
            L = 007
            Data: Ext = No extension
                  Ton = International
                  Npi = ISDN
```

```
                Address = 33970670001
        MAPPN_Modify_Registration_Status (521)
          L = 001
          Data: (1) :Deactivate
        MAPPN_Modify_Request_IP_SM_GW ellipsis (540)
          L = 078
          HALYS_ellipsis(0x01)        IP-SM-GW      Diameter      name
ipsmgw1.epc.mnc094.mcc208.3gppnetwork.org
          HALYS_ellipsis(0x02)        IP-SM-GW      Diameter      realm
epc.mnc094.mcc208.3gppnetwork.org
```

5.4.5.4 Registration of the reachability for SMS in the IP-SM-GW with MAP NOTE SUBSCRIBER DATA MODIFIED

When the HLR-HSS receives a MAP READY FOR SM from an SGSN or a NOTIFY Diameter from an MME, it can notify the IP-SM-GW with MAP.

```
    - - - - Super Detailed SS7 Analyser - - - - - - -
        PA_Len = 41
        MAP-OPEN-REQ (1)
          dest_address(Q713) (1)
          L = 013
          Data:  Route on SSN, Global Title included(0x04),
                 Signalling Point Code (ITU) = 1-201-1 ( 3657)
                 Subsystem Number = MSC (8),
                 Global Title :
                     Translation Type = 0,
                     Numbering Plan = ISDN/Telephony(E164),
                     Encoding Scheme = BCD, odd number of digits,
                     Nature Address Indicator = International number,
                     Address information = 33970670007
          orig_address(Q713) (3)
          L = 011
          Data:  Route on GT, Global Title included(0x04),
                 No SPC in address
                 Subsystem Number = HLR (6),
                 Global Title :
                     Translation Type = 0,
                     Numbering Plan = ISDN/Telephony(E164),
                     Encoding Scheme = BCD, odd number of digits,
                     Nature Address Indicator = International number,
                     Address information = 33970670006
        MAPorCAP_Application_Context (11)
          L = 009
          Data: (Hex) 060704000001001603
                    MAP_Subscriber_Data_Modification_Notification_V3
    - - - - - - - - - - - - - - - - - - - - - - - - - - -
----------------------- HEX ------------------------ [ 20170816
_142037_631175 ]
    - - - - Super Detailed SS7 Analyser (C)HALYS (Trace Level 8) - - -
        PA_Len = 38
```

```
MAP-NOTE-SUBS-DATA-MODIFIED-REQ (536)
  MAPPN_service_type (248)
    L = 002
    Data: (Hex) 0218
  MAPPN_timeout (45)
    L = 002
    Data: timeout value =  15 sec
  MAPPN_invoke_id (14)
    L = 001
    Data: 1
  MAPPN_imsi (18)
    L = 008
    Data: Address = 208940000001065
  MAPPN_msisdn (15)
    L = 007
    Data: Ext = No extension
          Ton = International
          Npi = ISDN
          Address = 33757791065
  MAPPN_UE_reachable (593)
    L = 001
    Data: (1) :sgsn
- - - - - - - - - - - - - - - - - -- - - - - - - - -
```

5.4.5.5 Simpler registration for SIP message reception (recommended)

A simpler compatible alternative implementation exists for simpler networks with a single SM-IP-GW. The ip_sm_gw_num (the GT of the IP_SM_GW) is a configuration parameter of the HLR-HSS. As illustrated in Figure 5.2, whenever it receives a SAR/Cx from the S-CSCF, the UE is then considered to be capable of receiving SMS-MT/SIP. In order to do so, the ip_sm_gw_num of the subscriber is assigned when the SAR is received containing the S-CSCF. It does not require the S-CSCF to have Sh/Diameter to obtain the IFC of the subscriber.

However, if the core IMS is fully 3GPP as in the previous method, this parameter will be overwritten by the gsmSCF address received in the ATM request.

5.5 Detailed procedure for SMS-MT and SMS-MO single segment

This section explains the details of SMS using Diameter between SMSC and the SMS-capable MME.

The procedure and even the payload format are very much as in the well-known MAP SMS procedure. The payload SM-RP-UI is simply encapsulated in the FORWARD_SHORT_MSG of the SGd/Diameter protocol between the IP-SM-GW and the UE through the MME and the eNodeB.

5.5.1 SMS-MT

From a 3G handset, an SMS is sent to a subscriber which is under a 4G coverage, its MME and SMSC (with an IP-SM-GW function is SMS capable. We give the detailed call flow and traces of the SMS-MT exchange between the SMSC and the UE through the MME. The 3G MAP SMS is classical and not shown. This is the simplest case of an SMS of less than 160 characters with "Test inactif6." In 4G SMS, there is the need as in MAP of segmenting SMS of more than 140 octets (see the detailed traces at the end).

5.5.2 SMS-MO

The UE 4G resends the preceding received SMS to the sending 3G subscriber. We describe only the 4G SMS-MO (see the traces at the end with the SMS status report).

5.6 Long SMS with segmentation

5.6.1 Long SMS-MT from 3G to a 4G coverage handset

From +33970675102 (3G) to +33970675066 (4G) with the below text (236 characters):

EXECUTIVE TELECOM is a French mobile operator (MVNO) whose clients are frequent travelers with very high ARPU and eager to have the best connections in situation of mobility, as well as best rates for data and voice during their travels.

The long SMS-MO received from 3G is reassembled and split into four pieces by the SMSC and then sent through the IWF-SMS 4G ↔ 3G converter with the SGd Diameter protocol (see the traces below). We show only segments 3 and 4 (segment 3 is sent by the SMSC before segment 4 as the sequence number will reorder them).

```
No.
    47 id-downlinkNASTransport, Downlink NAS transport(DTAP) (SMS)
CP-DATA (RP) RP-DATA (Network to MS)  (Short Message fragment 3 of 4)
```

```
Frame 47: 280 bytes on wire (2240 bits), 280 bytes captured
           (2240 bits)
Linux cooked capture
Internet Protocol Version 4, Src: 192.168.2.113 (192.168.2.113),
                             Dst: 192.168.2.117 (192.168.2.117)
Stream Control Transmission Protocol, Src Port: s1-control (36412),
               Dst Port: s1-control (36412)
S1 Application Protocol
    S1AP-PDU: initiatingMessage (0)
        initiatingMessage
            procedureCode: id-downlinkNASTransport (11)
            criticality: ignore (1)
            value
                DownlinkNASTransport
                    protocolIEs: 3 items
                        Item 0: id-MME-UE-S1AP-ID
                        Item 1: id-eNB-UE-S1AP-ID
                        Item 2: id-NAS-PDU
                            ProtocolIE-Field
                                id: id-NAS-PDU (26)
                                criticality: reject (0)
                                value
                                    NAS-PDU:
270faf7b47070762ae3901ab010307913379600700f7009f...
                                    Non-Access-Stratum (NAS)PDU
                                        0010 .... = Security header
type: Integrity protected and ciphered (2)
                                        .... 0111 = Protocol
discriminator: EPS mobility management messages (0x07)
                                        Message authentication code:
0x0faf7b47
                                        Sequence number: 7
                                        0000 .... = Security header
type: Plain NAS message, not security protected (0)
                                        .... 0111 = Protocol
discriminator: EPS mobility management messages (0x07)
                                        NAS EPS Mobility Management
Message Type: Downlink NAS transport (0x62)
                                            NAS message container
                                                Length: 174
                                                NAS message container
content: 3901ab010307913379600700f7009f640b913379605701f2...
                                                    GSM A-I/F DTAP -
                                                    CP-DATA
                                                    GSM A-I/F RP - RP-DATA
                                                    (Network to MS)
                                                    GSM SMS TPDU
                                                    (GSM 03.40)
                                                    SMS-DELIVER
                                                        0... .... = TP-RP:
TP Reply Path parameter is not set in this SMS SUBMIT/DELIVER
                                                        .1.. .... =
```

```
TP-UDHI: The beginning of the TP UD field contains a Header in
    addition to the short message
                                            ..1. .... =
TP-SRI: A status report shall be returned to the SME
                                            .... .1.. =
TP-MMS: No more messages are waiting for the MS in this SC
                                            .... ..00 =
TP-MTI: SMS-DELIVER (0)
                                            TP-Originating-
Address - (33970675102)
                                            Length: 11
address digits
                                            1... .... :
No extension
                                            .001 .... :
Type of number: (1) International
                                            .... 0001 :
Numbering plan: (1) ISDN/telephone (E.164/E.163)
                                            TP-OA Digits:
33970675102
                                            TP-PID: 0
                                            00.. .... :
defines formatting for subsequent bits
                                            ..0. .... :
no telematic interworking, but SME-to-SME protocol
                                            ...0 0000 :
the SM-AL protocol being used between the SME and the MS (0)
                                            TP-DCS: 8
                                            00.. .... =
Coding Group Bits: General Data Coding indication (0)
                                            00.. .... :
General Data Coding indication
                                            ..0. .... :
Text is not compressed
                                            ...0 .... :
Reserved, no message class
                                            .... 10.. :
Character set: UCS2 (16 bit)
                                            .... ..00 :
Message Class: Class 0 (reserved)
                                            TP-Service-Centre-
                                            Time-Stamp
                                                Year 17,
                                                Month 04,
                                                Day 27
                                                Hour 14,
                                                Minutes 07,
                                                Seconds 48
                                                Timezone:
                                                GMT + 2 hours
                                                0 minutes
                                            TP-User-Data-
                                            Length: (140)
```

```
depends on Data-Coding-Scheme
                                                TP-User-Data
                                                    User-Data
                                                    Header
                                                        User Data
                                                        Header
                                                        Length (5)
                                                        IE:
Concatenated short messages, 8-bit reference number (SMS Control)
                                                        Information
Element Identifier: 0x00
                                                        Length: 3
                                                        Message
                                                        identifier:
                                                        9
                                                        Message
                                                        parts: 4
                                                        Message
                                                        part
                                                        number: 3
                                                    [SMS text:
connections in situation of mobility, as well as best rates for data]
                                                Reassembled in: 65
No.
    65 id-downlinkNASTransport, Downlink NAS transport(DTAP) (SMS)
CP-DATA (RP) RP-DATA (Network to MS)   (Short Message Reassembled)
Frame 65: 212 bytes on wire (1696 bits), 212 bytes captured
(1696 bits)
Linux cooked capture
Internet Protocol Version 4, Src: 192.168.2.113 (192.168.2.113), Dst:
192.168.2.117 (192.168.2.117)
Stream Control Transmission Protocol, Src Port: s1-control (36412),
Dst Port: s1-control (36412)
S1 Application Protocol
    S1AP-PDU: initiatingMessage (0)
        initiatingMessage
            procedureCode: id-downlinkNASTransport (11)
            criticality: ignore (1)
            value
                DownlinkNASTransport
                    protocolIEs: 3 items
                        Item 0: id-MME-UE-S1AP-ID
                        Item 1: id-eNB-UE-S1AP-ID
                        Item 2: id-NAS-PDU
                            ProtocolIE-Field
                                id: id-NAS-PDU (26)
                                criticality: reject (0)
                                value
                                    NAS-PDU: 2785
    ad52c60907626e49016b010407913379600700f7005f...
                                Non-Access-Stratum (NAS)PDU
                                    0010 .... = Security header
type: Integrity protected and ciphered (2)
```

.... 0111 = Protocol discriminator: EPS mobility management messages (0x07)

Message authentication code: 0x85ad52c6

Sequence number: 9

0000 .... = Security header type: Plain NAS message, not security protected (0)

.... 0111 = Protocol discriminator: EPS mobility management messages (0x07)

NAS EPS Mobility Management Message Type: Downlink NAS transport (0x62)

NAS message container

Length: 110

NAS message container content: 49016b010407913379600700f7005f640b913379605701f2...

GSM A-I/F DTAP - CP-DATA

GSM A-I/F RP - RP-DATA (Network to MS)

Message Type RP-DATA (Network to MS)

RP-Message Reference

RP-Message Reference: 0x04 (4)

RP-Originator Address - (33970670007)

Length: 7

1... .... = Extension: No Extension

.001 .... = Type of number: International Number (0x01)

.... 0001 = Numbering plan identification: ISDN/Telephony Numbering (ITU-T Rec. E.164 / ITU-T Rec. E.163) (0x01)

BCD Digits: 33970670007

RP-Destination Address

Length: 0

RP-User Data

Length: 95

TPDU (not displayed)

GSM SMS TPDU (GSM 03.40)

SMS-DELIVER

0... .... = TP-RP: TP Reply Path parameter is not set in this SMS SUBMIT/DELIVER

.1.. .... =

```
TP-UDHI: The beginning of the TP UD field contains a Header in
addition to the short message
                                                ..1. .... =
TP-SRI: A status report shall be returned to the SME
                                                .... .1.. =
TP-MMS: No more messages are waiting for the MS in this SC
                                                .... ..00 =
                    TP-MTI: SMS-DELIVER (0)
                                            TP-Originating-
                                            Address -
                                            (33970675102)
                                                Length: 11
        address digits
                                                1... .... :
No extension
                                                .001 .... :
Type of number: (1) International
                                                .... 0001 :
Numbering plan: (1) ISDN/telephone (E.164/E.163)
                                            TP-OA Digits:
            33970675102
                                            TP-PID: 0
                                                00.. .... :
defines formatting for subsequent bits
                                                ..0. .... :
no telematic interworking, but SME-to-SME protocol
                                                ...0 0000 :
the SM-AL protocol being used between the SME and the MS (0)
                                            TP-DCS: 8
                                                00.. .... =
Coding Group Bits: General Data Coding indication (0)
                                                00.. .... :
General Data Coding indication
                                                ..0. .... :
Text is not compressed
                                                ...0 .... :
Reserved, no message class
                                                .... 10.. :
Character set: UCS2 (16 bit)
                                                .... ..00 :
Message Class: Class 0 (reserved)
                                            TP-Service-Centre-
                                            Time-Stamp
                                                Year 17,
                                                Month 04,
                                                Day 27
                                                Hour 14,
                                                Minutes 07,
                                                Seconds 50
                                                Timezone:
                                                GMT + 2 hours
                                                0 minutes
                                            TP-User-Data-
```

```
Length: (76) depends on Data-Coding-Scheme
                                              TP-User-Data
                                                  User-Data
                  Header
                                                      User Data
                  Header Length (5)
                                                      IE:
Concatenated short messages, 8-bit reference number (SMS Control)
                                                      Information
Element Identifier: 0x00
                                              Length: 3
                                              Message
                                              identifier:
                                              9
                                              Message
              parts: 4
                                              Message part
              number: 4
                                          [SMS text:
EXECUTIVE TELECOM is a French mobile operator (MVNO) whose clients are
frequent travellers with very high ARPU]
                                          [4 Short Message
fragments (472 bytes): #9 (134), #27 (134), #47 (134), #65 (70)]
                                              [Frame: 9,
payload: 0-133 (134 bytes)]
                                              [Frame: 27,
payload: 134-267 (134 bytes)]
                                              [Frame: 47,
payload: 268-401 (134 bytes)]
                                              [Frame: 65,
payload: 402-471 (70 bytes)]
                                              [Short Message
fragment count: 4]
                                              [Reassembled
Short Message length: 472]
```

5.6.2 The 4G resends (SMS-MO) the long SMS received from the 3G

From +33970675066 to +33970675102 with the previously received text (236 characters):

EXECUTIVE TELECOM is a French mobile operator (MVNO) whose clients are frequent travelers with very high ARPU and eager to have the best connections in situation of mobility, as well as best rates for data and voice during their travels.

Idem as for SMS-MT, the long SMS-MO is spilt into four segments with the SGd Diameter protocol. Only segment 4 is shown.

Figure 5.3 Interworking 3G ↔ 4G ↔ 3G for a long segmented SMS. The 3G handset received what it has sent. Screenshot of the sending 4G.

```
No.
     24 SACK id-uplinkNASTransport, Uplink NAS transport(DTAP) (SMS)
CP-DATA (RP) RP-DATA (MS to Network)  (Short Message Reassembled)

Frame 24: 244 bytes on wire (1952 bits), 244 bytes captured
           (1952 bits)
Linux cooked capture
Internet Protocol Version 4, Src: 192.168.2.117 (192.168.2.117), Dst:
192.168.2.113 (192.168.2.113)
Stream Control Transmission Protocol, Src Port: s1-control (36412),
Dst Port: s1-control (36412)
S1 Application Protocol
    S1AP-PDU: initiatingMessage (0)
        initiatingMessage
            procedureCode: id-uplinkNASTransport (13)
            criticality: ignore (1)
            value
                UplinkNASTransport
                    protocolIEs: 5 items
                        Item 0: id-MME-UE-S1AP-ID
                        Item 1: id-eNB-UE-S1AP-ID
                        Item 2: id-NAS-PDU
                            ProtocolIE-Field
                                id: id-NAS-PDU (26)
                                criticality: reject (0)
```

```
                                    value
                                 NAS-PDU:
        27335013931407636809016500080007918497903980f059...
                                 Non-Access-Stratum (NAS)PDU
                                 0010 .... = Security header
type: Integrity protected and ciphered (2)
                                 .... 0111 = Protocol
discriminator: EPS mobility management messages (0x07)
                                 Message authentication code:
           0x33501393
                                 Sequence number: 20
                                 0000 .... = Security
header type: Plain NAS message, not security protected (0)
                                 .... 0111 = Protocol
discriminator: EPS mobility management messages (0x07)
                                 NAS EPS Mobility Management
Message Type: Uplink NAS transport (0x63)
                                 NAS message container
                                 Length: 104
                                 NAS message container
content: 09016500080007918497903980f05961000b913379605701...
                                    GSM A-I/F DTAP -
CP-DATA
                                       Protocol
Discriminator: SMS messages
                                       .... 1001 =
Protocol discriminator: SMS messages (0x09)
                                       0... .... =
TI flag: allocated by sender
                                       .000 .... =
TIO: 0
                                    DTAP Short Message
Service Message Type: CP-DATA (0x01)
                                    CP-User Data
                                       Length: 101
                                       RPDU (not
                                       displayed)
                                    GSM A-I/F RP - RP-DATA
                                    (MS to Network)
                                    Message Type
                                    RP-DATA
                                    (MS to Network)
                                    RP-Message
                                    Reference
                                       RP-Message
                                       Reference:
                                       0x08 (8)
                                    RP-Originator
                                    Address
                                       Length: 0
                                    RP-Destination
Address - (48790993080)
                                       Length: 7
```

```
                                             1... .... =
Extension: No Extension
                                             .001 .... =
Type of number: International Number (0x01)
                                             .... 0001 =
Numbering plan identification: ISDN/Telephony Numbering (ITU-T Rec.
E.164 / ITU-T Rec. E.163) (0x01)
                                             BCD Digits:
48790993080
                                        RP-User Data
                                        Length: 89
                                        TPDU (not
                                        displayed)
                                  GSM SMS TPDU
                                  (GSM 03.40)
                                  SMS-SUBMIT
                                      0... .... =
TP-RP: TP Reply Path parameter is not set in this SMS SUBMIT/DELIVER
                                      .1.. .... =
TP-UDHI: The beginning of the TP UD field contains a Header in
    addition to the short message
                                      ..1. .... =
TP-SRR: A status report is requested
                                      ...0 0... =
TP-VPF: TP-VP field not present (0)
                                      .... .0.. =
TP-RD: Instruct SC to accept duplicates
                                      .... ..01 =
TP-MTI: SMS-SUBMIT (1)
                                  TP-MR: 0
                                  TP-Destination-
Address - (33970675102)
                                      Length: 11
              address digits
                                      1... .... :
No extension
                                      .001 .... :
Type of number: (1) International
                                      .... 0001 :
Numbering plan: (1) ISDN/telephone (E.164/E.163)
                                      TP-DA Digits:
33970675102
                                  TP-PID: 0
                                  00.. .... :
defines formatting for subsequent bits
                                      ..0. .... :
no telematic interworking, but SME-to-SME protocol
                                      ...0 0000 :
the SM-AL protocol being used between the SME and the MS (0)
                                  TP-DCS: 8
                                  00.. .... =
Coding Group Bits: General Data Coding indication (0)
                                      00.. .... :
```

```
General Data Coding indication
                                              ..0. .... :
Text is not compressed
                                              ...0 .... :
Reserved, no message class
                                              .... 10.. :
Character set: UCS2 (16 bit)
                                              .... ..00 :
Message Class: Class 0 (reserved)
                                       TP-User-Data-
                                       Length:
                     (76) depends on Data-Coding-Scheme
                                       TP-User-Data
                                         User-Data
                                         Header
                                             User Data
                                             Header
                                             Length (5)
                                             IE:
Concatenated short messages, 8-bit reference number (SMS Control)
                                             Information
Element Identifier: 0x00
                                             Length: 3
                                             Message
                                             identifier:
                                             47
                                             Message
                                             parts: 4
                                             Message
                                             part
                                             number:
                                             4
                                             [SMS text:
EXECUTIVE TELECOM is a French mobile operator (MVNO) whose clients are
frequent travellers with very high ARP]
                                       [4 Short Message
fragments (472 bytes): #8(134), #14(134), #19(134), #24(70)]
                                             [Frame: 8,
payload: 0-133 (134 bytes)]
                                             [Frame: 14,
payload: 134-267 (134 bytes)]
                                             [Frame: 19,
payload: 268-401 (134 bytes)]
                                             [Frame: 24,
payload: 402-471 (70 bytes)]
                                             [Short Message
fragment count: 4]
                                             [Reassembled
Short Message length: 472]
                     Item 3: id-EUTRAN-CGI
                     Item 4: id-TAI
```

5.7 Application to OTA SIM in pure PMR 4G networks

OTA SIM, SMS-based, commonly uses SMS-MT to update the SIM cards and SMS-MO to obtain the PoR. The IP-SG-GW allows any OTA server to operate in a pure 4G environment such as in certain PMR networks. To conclude, here is the result of updating the multi-IMSI SIM Tool Kit name (Figure 5.4). The format of the payload if BIP/CAT-TP or HTTPs OTA-IP is given in Section 12.

5.8 Mobile and fixed number portability with Dx/diameter to send SMS to IMS networks

5.8.1 LIR/Cx/diameter is the equivalent IMS of a legacy 3G MAP SEND_ROUTING_INFO req

LIR is the equivalent for voice calls of an SRI, not of an SRI_FOR_SM but the Server Name it returns gives the destination network before or after a ported-out case.

Figure 5.4   OTA SIM update "PCSTORM" of the SIM Tool Kit using 4G SMS.

Table 5.1 Cx/diameter, MAP, SIP and ISUP messages usable for number portability

	Message	Parameter to Specify the Destination	Invoking Network Entity
IMS (Diameter)	Location-Information-Request(LIR)	Public Identity AVP	I-CSCF (Origin-Host)
"	Location-Information-Answer(LIA)	Server-Name	
3G (MAP)	SEND_ROUTING_INFO req (SRI)	MSISDN	GMSC (GT)
"	SEND_ROUTING_INFO cnf	Roaming Number	
IMS (Diameter) voice call	SIP INVITE	MSISDN@Server-Name	S-CSCF
3G (MAP)	ISUP Initial Address Message(IAM)	Roaming Number	GMSC

5.8.2 Principle of the use of the location-information-request/Cx diameter to resolve the portability

The Cx interface is used when the destination-host is not known just some nominal destination-realm based on the public identity=MSISDN, which will be always the case for a first registration by the I-CSCF (if there are multiple HSSs) or for a ported-out number. There is only one difference with the MAP sending case: how the portability is resolved in the receiving country side (at Level $N - 1$) as we exclude the case that there will be a practical case of MNP resolution at level $N - 2$ (see [0.7] for the terms) in the originating side.

For the implementation in the destination country, it is interesting to note that it cannot be done by the I-CSCF (the IMS equivalent of a GMSC MAP) with *a portability database* as *this is not part of the I-CSCF standard*. The LIR will go to the nominal (Hong Kong CSL) HSS. The LIA answer will give a "Server Name" which is Hutchison HK. The requesting entity, SMSC or I-CSCF, will then send the next request to the Hutchison HK Server Name:

- MT-Forward-SM for SMS-MT.
- SIP INVITE for an outgoing call (see Figure 5.5).

Figure 5.5 shows a test tool which simulates the I-CSCF and sends a Dx/Diameter commands (an LIR with "Public Identity" = 852604123456@mnc000.mcc454.3gppnetworks.org, which is a range of the MNO Hong-Kong CSL) and gets the LIA. The Server Name returned in

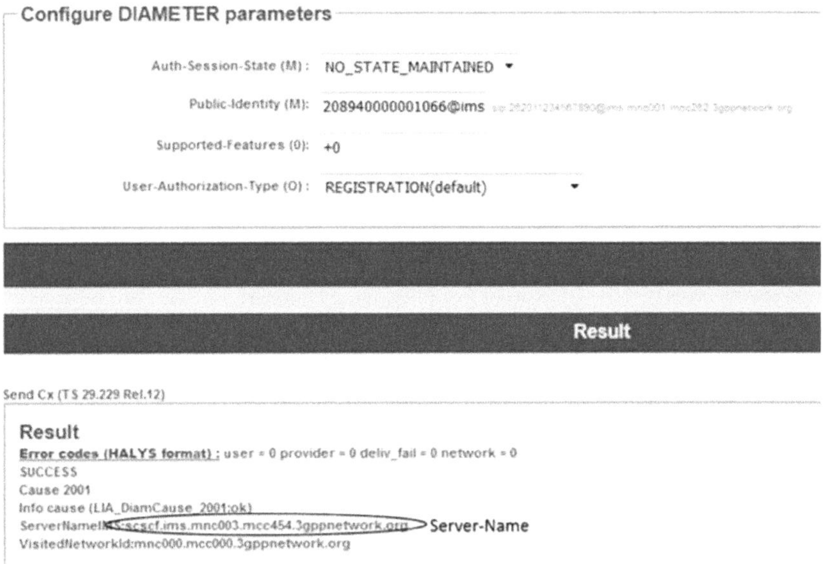

Configure DIAMETER parameters

Auth-Session-State (M) :	NO_STATE_MAINTAINED ▾
Public-Identity (M):	208940000001066@ims ⠀⠀⠀⠀⠀⠀⠀⠀⠀
Supported-Features (O):	+0
User-Authorization-Type (O) :	REGISTRATION(default) ▾

Result

Send Cx (TS 29.229 Rel.12)

Result
Error codes (HALYS format) : user = 0 provider = 0 deliv_fail = 0 network = 0
SUCCESS
Cause 2001
Info cause (LIA_DiamCause_2001;ok)
ServerName scscf.ims.mnc003.mcc454.3gppnetwork.org → Server-Name
VisitedNetworkId:mnc000.mcc000.3gppnetwork.org

Figure 5.5 Interrogating an HSS-SLF with location information Req/Dx to resolve the number portability.

the LIA is "scscf.mnc003.mcc454.3gppnetwork.org" which shows that the number has been ported-out to Hutchison Hong-Kong.

5.8.3 Fixed ↔ mobile portability

The Server Name can also be used directly to send SMS, with an MT-Forward-SMS to Hong-Kong CSL. It needs not to be an MNO Server-Name with VoLTE and IMS ".3gppnetwork.org." It can be from a fixed terminal Internet network with a standard core IMS such as ims.ipnexia.,be, ims.halys.fr, etc. So that for calls and SMS, the HSS will also resolve the portability of numbers fixed to mobile and mobile to fixed. Also note that the number of a given country could be ported to another country without any technical difficulty.

5.8.4 How to implement the portability of a number in the ported-out network

The record *must be kept in the HSS* after porting-out. When Hong-Kong CSL is informed that +852604123456 is ported out to Hutchison HK, they just

Figure 5.6   MMS interworking: 3G-4G UE MMS → received by the SIP destination client.

enter the new Server-Name in the HSS record "scscf.mnc003.mcc454.3gpp
network.org." If the subscriber returns to Hong-Kong CSL after having tried
the other networks, they just replace the Server-Name by their own to avoid
loops.

5.9 3G ↔ SIP MMS interworking

In Figure 5.2 (not shown for simplicity), the SMSC communicates with the
MMSC of the MNO and receives the MMS notification SMS when an MMS-
MT is sent out to one of the MNO's subscribers from the MNO's MMSC.
The MMSC is a data service accessible through the PGW-GGSN.

  The 3G ↔ SIP MMS interworking described below is a clever original
feature beyond the IP-SM-GW function, compatible with all the 3GPP standards referred at the end of this chapter. It enhances the SIP → legacy network
convergence to the multimedia messages.

5.9.1 SIP receiving of 3G MMS

When an MMS is sent to the MNO's subscriber whether it is registered SIP or using the 3G or 4G network, the MNO's MMSC receives it (an MM1 m-send-req if it is from a subscriber of the same network or an MM4 Forward-req if it comes from another network with MMS interworking enabled). In both cases, the received *MMS content is first stored in the MNO's MMSC* for delivery.

The MMSC must send a notification by SMS-MT to the destination and does not know which network or SIP it is using. It interrogates the HLR-HSS and obtains the information, and then knows that it is only registered SIP (no 3G MMS). It then converts the content to an HTML format if it is not 3G enabled instead of an MM1 or MM4 format [5.8]. The "SMS notification" for the MMS on the left of Figure 5.6 refers to the *HTML format ".htm" of a page stored in the MMSC* which is received by the SIP client as a MESSAGE in the "Chat" menu.

The user simply clicks on the URL and receives the picture (right of Figure 5.6) as this URL points to the MNO's MMSC storage which is addressable from the UEs as they both belong to the MNO. Figure 5.6 shows an MMS sent by a subscriber using 3G to a SIP registered subscriber of the same HPLMN.

5.9.2 Sending an MMS from the SIP client to a 3G UE

The SIP client sends the "MMS" as an attachment in a Chat MESSAGE. If the destination is registered in a 3G network, the SIP MESSAGE goes in any case through the IP-SM-GW, which recognizes (a content analysis which shows http) that it must be sent as "MMS" to the destination. It behaves as the destination UE and 1) *retrieves the attachment* from the sending SIP client and then 2) *sends it to the MMSC* with a simulated MM1 m-send-req.

The sending of the MMS to the destination by the MMSC is then the normal MMS procedure [3.8]: MM1 or MM4.

References

[5.1] 3GPP TS 29.002 V14.3.0 (2017-05), "Digital cellular telecommunications system (Phase 2+) (GSM); Universal Mobile Telecommunications System (UMTS); Mobile Application Part (MAP) specification," Release 14. *Updated MAP with the support of IP-SM-GW notably.*

[5.2] 3GPP TS 29.338 V14.1.0 (2017-03), "Universal Mobile Telecommunications System (UMTS); LTE; Diameter based protocols to support

Short Message Service (SMS) capable Mobile Management Entities (MMEs)," Release 14.

[5.3] 3GPP TS 29.228 v14.2.0 (2017-03), "Digital cellular telecommunications system (Phase 2+); Universal Mobile Telecommunications System (UMTS); LTE; IP Multimedia (IM) Subsystem Cx and Dx Interfaces; Signalling flows and message contents," *Communications between HSS et IMS.Annex B gives the XML format of the User Profile in the AVP User-Data.*

[5.4] 3GPP TS 29.229 v14.1.0 (2017-10), "Digital cellular telecommunications system (Phase 2+); Universal Mobile Telecommunications System (UMTS); LTE; Cx and Dx interfaces based on the Diameter protocol; Protocol details," Release 12, *Description of the Diameter protocol messages between HSS and I-CSCF of IMS.*

[5.5] 3GPP TS 23.228 V14.3.0 (2017-05), "Digital cellular telecommunications system (Phase 2+) (GSM); Universal Mobile Telecommunications System (UMTS); LTE; IP Multimedia Subsystem (IMS); Stage 2," *The overall reference for the IMS architecture with all the IMS acronyms, interfaces Sh and Si (legacy for charging).*

[5.6] 3GPP TS 23.204 V14.0.0 (2017-04), "Universal Mobile Telecommunications System (UMTS); LTE; Support of Short Message Service (SMS) over generic 3GPP Internet Protocol (IP) access; Stage 2." *Standard IP-SM-GW architecture to interconnect SIP MESSAGE with 3G and 4G SMS.*

[5.7] 3GPP TS 23.038 V14.0.0 (2017-04), "Digital cellular telecommunications system (Phase 2+) (GSM); Universal Mobile Telecommunications System (UMTS); LTE; Alphabets and language-specific information." *Includes the GSM 7 bits alphabet and its translation to ISO 8859-1.*

[5.8] 3GPP TS 23.140 V6.16.0 (2009-4), "Digital cellular telecommunications system (Phase 2+); Universal Mobile Telecommunications System (UMTS); Multimedia Messaging Service (MMS); Functional description," *The MMS service still exists in 2017 and is quite convenient!*

[5.9] 3GPP TS 24.341 V14.0.0.(2017-04) "Digital cellular telecommunications system (Phase 2+) (GSM); Universal Mobile Telecommunications System (UMTS); LTE; Support of SMS over IP networks; Stage 3."

[5.10] ETSI TS 102 226 V13.0.0 (2016-05), "Smart Cards; Remote APDU structure for UICC based applications (Release 13)," SCP03.

6

Multicast: MCPTT PMR, MOOC Teaching, and TV in Local Loop Networks (RTTH)

..The countries which have the best performance for the Pisa rating are also those where the daily time spent on the Internet is the smallest and where children least use a computer for their home school work. The higher managers of the Silicon Valley massively choose for their own children selective schools where screens are excluded.
    Natacha Polony, Le Figaro, 26 August 2016.

6.1 Operational need for multicast in PMR networks

Using IP and with the overhead, the voice service of PMR requires about 50 Kb/s; the good quality video sharing application (mp4 codec) requires 2 Mb/s; an eNodeB using the PPDR band 2×3 Mkz in the 700-MHz sideband used in certain countries offers 17 Mb/s DL close to the antenna, not more than 3G. An economical satellite backhaul link has 128 Kb/s which is enough for signaling but can at most handle two voice 50-kb/s channels.

This means that for the telephone service when all security forces are close to the antenna, one could rely on unicast channels up to about 340 members (17/0.05), if they are all at the coverage limit, the rate is 1/4 that is then 85 members which is a good size group and there is no need for multicasting the downlink. We provide below a better average value assuring a uniform distribution of the security forces in a circle limited by the coverage.

6.2 Triple play, the need for multicast TV and massive open online course (MOOC)

A commercial 4G eNodeB (for example, Band 42 used for local loop) provides 70 Mb of downlink bandwidth with 15 MHz.

The Triple play service is Internet + voice + TV. For TV broadcast, the efficient new H265/HeVC video codec which require only 2 Mb/s per TV channel but is not supported by all devices and many use H264/Mpeg-4 which needs the double for the same quality. The 4G local network is a cost-effective alternative to fiber cables pulled to all homes; it can also provide the TV service (which in "white areas" will be very poor using radio broadcasting). If Mpeg-4 is used and one does not want to devote more than 20 Mb/s for the TV service, there could be only five viewers using the traditional unicast method (two persons watching the same channel use their own 4G channel). This means that 4G TV using unicast is not the right solution. If the 4G Multicast is used instead, five programs could be proposed without any limitation on the number of viewers.

There is then the delicate question of selecting which five programs to multicast as none other could be broadcasted without degrading the Internet and telephony primary service. Some could think that this limitation is useful as it forces to arbitrate for the most useful programs: education and culture and drop most music and game shows (no hesitation for the local versions of "guess my age," "guess the words," the "weak link," as well as all the interviews of talk shows). The news only channels are not needed as two of the five channels would provide the news service among their other programs at 12H and 20H only. One of the channels of would be purely allocated to education: Maths, Physics, and Chemistry level license, geography, history, literature, and foreign languages. As recreation on this channel, after 22 PM, "l'instit" with the excellent Gérard Klein or Physics conferences by Etienne Klein, because he is an excellent lecturer, not because he has the same name as the actor or because he gives a bad example of much copying texts from other authors. It is certain that the children raised in this happy 4G local loop region (watching the few films such as "the 7th seal (an incentive to become a good chess player who can beat the Death in the final scene)," "il est minuit Dr. Schweitzer (to help his fellow man)," "Potemkine (revolutionary preparation)," and "Urga (for the beauty and Chatrov's 'on the hills of Mandchuria' valtz)," would be a reward of their good school work; they would be very successful throughout their education and would also become politically smart and useful citizens, as well as their parents.

How to finance these 4G local loop networks, which are likely to depend on local investments? There are politically two traditional approaches:

- Taxing those which have successful children (the French "impôt sur la fortune"). The families whose children get A or B marks would be taxed, quite happily as some may enjoy looking down on their less lucky neighbors. This is the successful principle of the ISF tax in France which many of this tax payers would regret to see being suppressed as it is a standing sign. However, it creates jealousies and is not then the best solution for social peace.

- Taxing those which have lesser grades of fail. It was like when the Arab-Muslims occupied Palestine in the 630 AD years. They imposed the moderate "djizia" tax on the non-Muslims (that is mostly the other inhabitants) on the pretense of the protection and a dispense of the military service. This was a successful smooth method, and after two generations, many of the Palestine's Jews had peacefully converted to Islam giving descent to great men such as the well-known scientists Averroes and Avicenne which ascents had migrated to Morocco and then to Spain. After some years, every child would have at least B under the pressure of their parents.

What about the competing satellite TV? Of course, the authorities financing the 4G Multicast TV would take local regulations to forbid them for 1) their ungainly appearance and 2) their potential health danger even if they are receivers, the counsel doctor of the administration having stated that "their side auto-radiated spectrum effect on human tissues must be avoided."

We have then shown that this 4G local loop TV is most useful and that the economists had several financing alternatives. This is not the engineers' responsibility to discuss, but here is below the technical explanation of the 4G Multicast.

A case of use of commercial TV over 4G is often mentioned. The spectators at Roland Garros are seated watching a game, but they are also interested by a simultaneous side game from time to time. Unicast TV broadcasted to several hundred viewers would saturate the 4G: the solution is to set up a few multicast channels for all the simultaneous games. The same idea could be used for soccer; during a game, the spectators could have subscribed to a multicast TV, with an alert every time a goal is scored in one of the other simultaneous games. Another improvement idea is based on the fact that many are interested in watching only the goals. So, all goal sequences will be dispatched in multicast to all; they will be stored in the large cache of modern handsets, ready to be viewed immediately when selected. Two MNOs (Verizon and Telstra) are known to have launched Multicast in their network and it is expected that the number will increase in view of the large number of usage of video streaming.

It provides a much better quality video than the clever setup using satellite TV but with the need to procure the cheap multicast receivers; a low-bandwidth UL channel for the interactivity is used but it is the same idea.

6.3 Quantitative elementary modeling of the fiber vs 4G local loop choice

Intuitively, if there is a low density area, it is cheaper to provide the Internet coverage with a 4G local loop and an eNodeB station than installing a fiber to each household, N in total. Below, let us make a simple decision model. We will estimate the mean distance from the center to each user assuming their uniform distribution.

We model the area of each household by a hexagon so that they cover the whole 4G eNodeB coverage; this type of hexagonal "meshing" is quite usual to discretize the plane. Each hexagon below is an individual household all with the same size to simplify (Figure 6.1).

The distances, using as unit the distance between the centers of two adjacent hexagons, and the number of households from the center are:

distance: 0 1   2   3   4   5   6...
n 0   1   2   3   4   5...
households 1 6 12 24   48...   (geometric progression of ratio $r = 2$,
                                              total $N = 91$ for $n = 4$)

If there are N households in the maximum 4G coverage footprint, we compute the maximum distance n by solving the equation:

$$N = 1 + 6(2^0 + 2^1 + 2^2 + 2^3 ... + 2^n) \text{ that is}$$
$$N = 1 + (6(2^{n+1} - 1)/(2 - 1)$$
$$2^{n+1} = 1 + (N - 1)/6$$
$$n + 1 = \log_2(1 + (N - 1)/6) \tag{6.1}$$

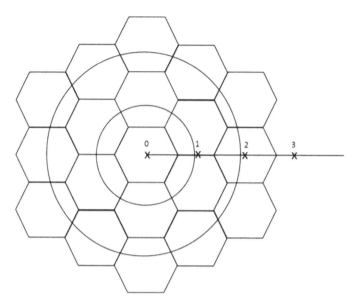

Figure 6.1 Distribution model of users evenly distributed in a low density area with a hexagonal meshing discretization.

Example:

$$N = 19 \text{ households} \rightarrow 2^{n+1} = 2 \rightarrow n = 1$$
$$N = 91 \text{ households} \rightarrow 2^{n+1} = 16 \rightarrow n = 3$$
$$N = 381 \text{ households} \rightarrow 2^{n+1} = (\log 64.333/\log 2) - 1;$$
$$n = 5.007 \text{ (not an integer in general)}$$

6.3.1 Average distance center – household with fibering

$$D = 0 \times 1 + 1 \times 6 + 2 \times 12 + 3 \times 24 + 4 \times 48 + 5 \times 96 \ldots$$
$$= 6(1 + 2 \times 2^1 + 3 \times 2^2 + 4 \times 2^3 + 5 \times 2^4 + \ldots)/N$$

(we have also $N = 91$ with $n = 3$). The sum is obtained by taking the derivative of the sum of the continuous series $u_m(x) = x^m$, which is $S_n(x) = 1 + 6(x^{n+1} - 1)/(x - 1)$, with n computed from (6.1).

This gives by derivation of $S_n(x)$:

$$D \times N = 6 \times dS_n(x)/dx$$
$$= 6 \times ((1 - x^{n+1})/(1 - x) - (n + 1)x^{n+1})/(1 - x), \quad (6.2)$$

that is, with $x = 2$ the geometric rate of the series (6.1):

$$D \times N = 6 \times ((1 - 2^{n+1}) + (n + 1)2^{n+1}) \quad (6.3)$$

This formula (6.3) gives the total number of fiber lengths to cover all households assuming their uniform distribution; one enters $n + 1$ computed with (6.1). This simplifies to a formula with N only (6 is the number of vertices for the hexagon used for meshing the area).

$$D \times N = 1 - N + (\log_2(1 + (N - 1)/6)) \times (1 + (N - 1)/6) \quad (6.4)$$

Example (Table 6.1):

This is a sketch of a method which allows deciding which is the best solution between installing one and several eNodeBs or fibering all households from a central dispatching point in the center of the concentration of households.

6.3.2 Cost model for the fibering solution to the home (FTTH) vs 4G radio (RTTH)

We have the FTTH data from a survey of major FTTH suppliers, mainly civil construction or cable pulling; the cost of the fiber cables is itself negligible (Table 6.2).

Table 6.1 Computation of the total length of fiber cabling for various numbers of households

Number of Households	Max Distance $= n + 1$	$D \times N$ Total Length of Point-to-Point Fibers	D Average Length of Point-to-Point Fiber to a Household
19	2	30	1.5789
91	4	294	3.2308
381	6.007	1924	5.0499
887	7.216	5515	6.2187
2000	8.384	14802	7.4010
4000	9.382	33545	8.3865
6139	10	55302	9.0083
12000	10.966	119590	9.9659

Table 6.2   Cost model for the fiber to the home (FTTH)

Situation	Average Cost/Meter	Rental of Supports or Pipes	Proportion of this Situation
Urban (trenches in street)	70 euros/m		10%
Urban (use of existing tubing from the incumbent operator)			90%
Country side (trenches)	30 euros/m		10%
Country side (use of PTT or electrical poles)		0.7 euros/m/year	70%
Country side (use of existing tubing)		???	20%
Country side	30 euros/m	??	10%
IRU (unrevocable right of usage)			Negligible

In the country side, it is observed that an eNodeB with a power of 40 dB can cover up to 10 km in the direction of one of its antenna sectors. So, the 4G local loop can cover a wide area. However, only 70 Mbits/s are available and it is appropriate to limit to 30 users for each sector to have an acceptable QoS for each of them. More generally with various eNodeB models, we have the estimates below for planning (Table 6.3).

Close to the station 256QAM may be used from Release 12 (8 bits/symbol), then 64QAM (6 bits), 16QAM (4 bits), and then QPSK (2 bits/symbol).

Several bandwidths may be available from the regulator, 10, 20 (as in 4G mobile), or 40 MHz in Band 42 (3.5 MHz) with an important effect of the investment. With 40 MHz, 100 subscribers could be supplied with 5 MHz of bandwidth, the minimum required by most regulators, that is less than half with RTTH than with the fibering FTTH. In most cases, even if FTTH provides more bandwidth, it is not really needed for Home and RTTH which is quick to deploy will remain for economic reasons. If a mast must be installed, the price with the secured room, power supply, and fence is 50–100 keuros.

Table 6.3   Cost model for the radio to the home (RTTH)

Number of Transmitters–Receivers	Power (Tri-Sectored eNodeB) (Watts)	Range (km)	Price of eNodeB (2018) with qty Discount (Euros)	Installation Price on an Existing Mast (keuros)
2T 2R	2 × 1	< 0.5	< 5k	10–15
2T 2R	2 × 10	3–5	< 10k	10–15
2T 2R	2 × 20	< 10	10–15k	10–15
4T 4R	4 × 20	10–15	15–20k	10–15
8T 8R	8 × 16	< 20	>25k	10–15

6.4 3GPP multicast architecture

Several implementations, such as illustrated in Figure 6.2, conveniently integrate the MBMS GW in the PGW as it also uses the GTPv2 protocol. This section and the following two explain the *first part: the establishment of MBMS bearers* for the multicast delivery and the joining of a multicast bearer by a UE using the GC1 unicast bearer with the AS.

To deliver contents to a UE, the AS on the right has two choices: unicast through the usual SGi interface and multicast through the SGi–mb interface. The UE and the AS interact through the GC1 interface, a normal unicast bearer (GC1 is a proprietary application protocol between the UE and the AS; it was supposed to have a standardization in Rel 13). Example of use: the UE reports the bad coverage for a unicast bearer and requests the creation of a multicast bearer while transmitting its location and the received signal level, or the contrary.

GC1 is also used by the UE application to request the temporary mobile group identity (TMGI) and flow identifier of the available multicast bearers, so it can attach to the reception of one of the corresponding MCCHs.

6.5 Detailed call flow of an MBMS session

An MBMS service area is a small geographical area covered by a multicast service.

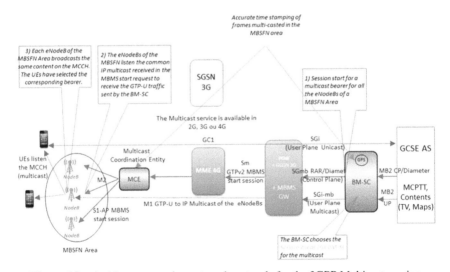

Figure 6.2   Architecture, equipment, and protocols for the 3GPP Multicast service.

6.5.1 Overall call flow

A given eNodeB initializes (M2 setup) the multicast service by sending to the MCE it depends on a list of service areas. When it receives it, the MCE relays this list to the MCE it depends on.

To initiate a multicast session, there is a list of service areas which may be over several MMEs (Figure 6.3). The MBMS GW has a map and is able to select the MMEs which cover them. The MBMS GW selects the MMEs for each service area from that list and sends an MBMS session start containing a sublist of service areas to each MME.

For each MCE, the MME extracts the intersection of the list received from the MBMS GW and the list received from each MCE and sends it to each MCE in an M3 session start.

The full traces of the MBMS bearer setup are in the Appendix to be used for training.

Below is the list of commands used by MCPTT AS (also called CGS AS), and then the BM-SC commands to start Multicast sessions, Stop, and Update. It must be noted that concerning the sending of data to the eNodeBs, *IP Multicast* was never used although it is in the specs, it is *IP Broadcast* (all eNodeBs which have MBMS sessions listen), and hence the command AAR from the MBMS GW to the BM-SC was never used (only the *Broadcast mode* of [6.10] is implemented). To implement the tactical network federation mechanism of Chapter 2, the "Notify New Network" is added. In Table 6.4, the commands not used in the current IP broadcast mode are in red.

The CGS AS manages the geographical mapping; *it has the map of all SAIs* and can select in the list corresponding to the area to be covered by the MBMS service of a list of groups TMGI. The CGS AS can select the BM-SC which covers a group of SAIs, eventually *several BM-SCs* per CGS AS to cover all the SAIs.

The BM-SC has the SAI → MBMS-GW mapping which provides the list of SAIs of an area which the addressed MBMS-GW covers (there may be *several MBMS-Gws per BM-SC*). The command to manage an MBMS service is then Re-Auth Request. The parameters TMGI, List of SAIs, and QoS are set by the CGS AS and conveyed down to the eNodeB, the BM-SC IP Address and Ports to be used for the user data MB2-U sent by the CGS AS to the BM-SC for the group identified by the TMGI parameter. For *the multicast user plane,* the data are sent to the BM-SC Address and then by the BM-SC to the IP Multicast Address of the group to be received by the eNodeBs of the service area.

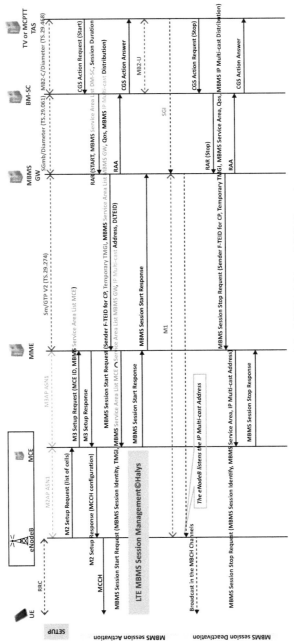

Figure 6.3   Call flow of the 3GPP Multicast MBMS service.

Table 6.4 MB2-C/Diameter messages CGS AS ↔ BM-SC

MB2-CP Commands TS 29.468	Short Name and Direction	Parameter and Equivalence
CGS Action Request START, STOP, UPDATE, HEARTBEAT	GAR CGS AS → BM-SC	MBMS StartStop-Indication TMGI Allocation Requested (*list of groups TMGI*) MBMS Flow Identifier (*for Update of a particular Flow*) MBMS Bearer Request (*list up to 256 Service Area Identifiers (SAI)*) QoS-Information (*Max Bandwidth, ARP*)
CGS Action Answer	GAA CGS AS ← BM-SC	TMGI Allocation Response (list of TMGIs successfully allocated) MBMS Bearer Response (list of SAIs successfully allocated) BM-SC Address (IP Broadcast where the BM-SC wants to receive the user data from the MB2-U interface with the GCS AS for that group). BM-SC Port (Port where the BM-SC wants to receive the user data from the MB2-U interface with the GCS AS for that group)
GCS Notification Request	GNR GS AS ← BM-SC	TMGI-Expiry MBMS-Bearer-Event-Notification
GCS Notification Response	GNA CGS AS → BM-SC	

—- MB2-U → — Sgi-mb ———→ —————— M1 —————→ —
MBCH →
   GCS AS ————BM-SC————→ MBMS-GW —————→
eNodeB(s) ———→
   UE———————————→ ——→

The BM-SC will then send RAR commands to the various MBMS-GW to create the various MBMS bearers (Table 6.5).

The *NNR* message is broadcasted to an IP broadcast address and then received by all IP-connected tactical networks. They return to the MBMS-GW of the joining tactical network their own MBMS-GW IP, so that the joining tactical network may add their service area for its group. This includes sending a Re-Auth Request command to the MBMS-GW of all the other

Table 6.5 SGmb/Diameter messages BM-SC ↔ MBMS GW

SGmb Commands TS 29.061	Short Name and Direction	Parameter and Equivalence
AA Request	AAR GGSN → BM-SC	Not used (only for IP Multicast). In this mode, the UE would have needed to register.
AA Answer	AAA GGSN ← BM-SC	Not used
Stop Request	STR GGSN → BM-SC	Not used (only for IP Multicast)
Stop Answer	STA GGSN ← BM-SC	Not Used
Re-Auth Request START, STOP, UPDATE, HEARTBEAT User data BM-SC → MBMS-GW uses unicast (with a Port number) User data MBMS-GW → eNodeBs use multicast	**RAR** BM-SC → MBMS-GW	MBMS StartSTop-Indication MBMS Service Area (*list of SAIs*) **MBMS Access Indicator** (*E-UTRAN*) **MBMS Session Duration** (*sec*) TMGI (*the Identity of each Group*) QoS-Information (*Max Bandwidth,ARP*) MBMS Time-to-Data Transfer (*sec*) MBMS User-Data-Mode-Indication (*Unicast*) CN-IP Multicast-Distribution (*Multicast*)
Re-Auth Answer (provides the BM-SC with the IP and Port to which the user data must be sent).	**RAA** BM-SC←MBMS-GW	MBMS GGSN-Address (contains the IP for the user plane data. MBMS-User-Data-Mode-Indication (unicast and multicast(1) indicates that BM-SC must use multicast) MBMS-GW-UDP-Port (UDP port from which the user plate date will be received by the MBMS-GW)
Notify New network Request	Broadcast to all networks #i **NNR** DRA#3→ BM-SC#i	IP new MMBS-GW #3
Notify New network Answer	**NNA** BM-SC#3 ← BM-SC#i	IP MBMS-GW#i

tactical networks. To create start MBMS sessions in all ot them for the group of the joining tactical network.

The MBMS-GW can in turn address *several MMEs* with a GTPv2 MBMS Session Start; they are not detailed here. They are received by the Sm interface of the MMEs which communicate using the M3/ASN1 protocol with an MCE.

Between the MME and the MCE, Table 6.6 lists all the messages.

6.5.2 M3/diameter messages MCE ↔ MME: Role of the MCE

M3 does not convey radio configuration data (it is "frequency agnostic"); it is purely logical. The MCE's role is to assign them to the eNodeB using the M2 protocol that includes frequency, modulation, and coding schemes. It also decides which transmission mode is used: MBMS or SC-PTM. The MCE may be centralized or distributed and is the scheduling entity for the multicast service, which means that it can reassign bandwidth for the various groups. In Chapter 2 for the federation of tactical networks, we have used a

Table 6.6   M3/ASN1 main messages

M3 Commands TS 36.344	Direction	Parameter and Equivalence
MBMS Session Start	eNodeB← MCE	MBMS e-RAB QoS MBMS Service Area (*list of SAIs*) MBMS Session Duration TMGI (*the Identity of each Group*) MBMS Minimum Time-to-Data Transfer (same as TS 29.061) IP Multicast Address IP Source Address GTP DL TEID
MBMS Session Start Response Session Start Failure	MCE → MME	
MBMS Session Update Request	MCE ← MME	
MBMS Session Update Response	MCE → MME	
M3 Setup Request	MCE → MME	
M3 Setup Response M3 Setup Failure	MCE ← MME	
MCE Configuration Update	MCE → MME	MBMS Service Area List
MCE Configuration Ack	MCE ← MME	

mixed architecture with the distributed MCEs coordinating for the multicast service, playing the same role as the PGW for the normal data service with pre-emptive priorities.

Table 6.7   M2/ASN1 main messages

M2 Commands TS 36.443	Short Name and Direction	Parameter and Equivalence
M2 SETUP Request	eNodeB → MCE	Global eNodeB ID eNodeB MBMS Configuration data Items (MBMS Service Area Items)
M2 SETUP Response	eNodeB ← MCE	MCCH related BCCH Configuration data Item, one per MBSFN Service Area (used by the eNodeB to configure the MCCH of each Service Area, with the RRC message *MBFSN Area Configuration* [6.10]
MBMS Session Start Allocated by the MBMS-GW	eNodeB ← MME	MBMS e-RAB QoS MBMS Service Area *(list of SAIs)* MBMS Session Duration TMGI *(the Identity of each Group)* MBMS Minimum Time-to-Data Transfer (same as TS 29.061) IP Multicast Address IP Source Address GTP DL TEID
MBMS Session Start Response MBMS Session Start Failure	eNodeB → MCE	
MBMS Scheduling Information	eNodeB ← MCE	PMCH Configuration MBMS Session List per PMCH MBSFN Subframe configuration Common Subframe Allocation Period MBSFN Area ID SFN MBMS Session Identity
MBMS Scheduling Information Response	eNodeB → MCE	

6.5.3 M2/diameter messages eNodeB ↔ MCE

An eNodeB is controlled by a single MCE which allocates the frequency, modulation, and coding schemes for the various multicast channels. The list of all messages between the eNodeB and the MCE is given by Table 6.7.

6.5.4 "Joining" (MBMS multicast activation by the user) GC1 UE → application server

After the creation of the MBMS bearer, all the concerned eNodeBs are set in a multicasting state with the data being multicast as soon a member has joined the multicast group. Any UE under coverage may request through its client GC1 (still proprietary in Rel 14 [6.7]) from the AS to join a list of multicast channels, which are returned with: their TMGI × flow identifiers, the IP Multicast, and a commercial name. The user can then select to receive a particular MBMS bearer and become a member of that multicast group. If the selected channel is not yet opened, there is a request to the AS to create it with a RAR/SGmb.

6.6 Centralized or distributed multicast coordination entity (MCE)

Figure 6.4 shows a centralized MCE in a VM environment. If there are several MBFSN areas, the allocation of MCCH channels requires a cooperation as several groups may have intersecting MBFSN and an eNodeB may have one MCCH channel for one group and one for the other.

6.7 MBMS delivery and eMBMS-capable device stack

There are four delivery modes applicable to various types of multicast services

- File download delivery
- Streaming delivery
- Group communication (mandatory for MCPTT)
- Transparent delivery.

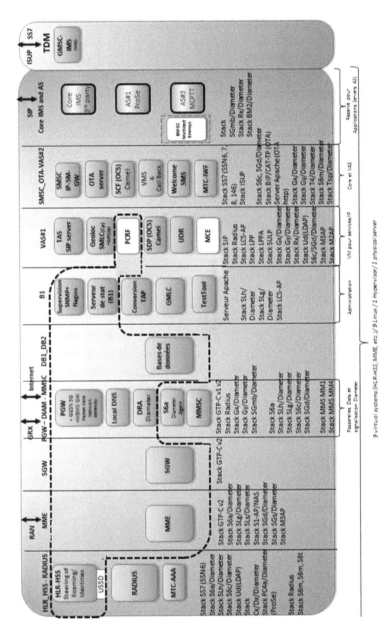

Figure 6.4   Reduced EPC for a 4G local loop network with multicast optional.

6.7.1 Group communication delivery appeared in [6.7, Rel 13]

This is the only one which does not require a "middleware" (such as provided by Expway) in the handset. This "black pipe" mode/UDP is specified for the MCPTT service.

6.7.2 Transparent delivery appeared in [6.7 Rel 14] and other modes

This is the latest mode included in the 2017 release of TS 23.246. The other modes such as Download delivery (used for Mobile TV) and Streaming Delivery (used for SW updates) appeared several years ago in [6.7 Rel 9].

6.8 Interoperability: Intergroup and interagency communication

Intergroup means that communication between several tactical networks is possible; this is done at the MCPTT AS level, if the backhaul link of the tactical networks is enabled.

Interagency communication with links between their MCPPT AS.

6.9 Architecture with virtual machines

In Figure 6.4, the MCE had been included in one of the VMs. A separate MCE is something relatively costly (there are few companies providing one) and it is simpler to install it as SW on one of the VMs of the core network.

The eNodeBs (S1–AP interface) must be able to reach simultaneously the MME IP address [VM#2] and their M2 interface to reach the MCE IP (VM#7).

What is inside the dashed zone is enough for a 4G EPC local loop such as used as an alternative to provide high-bandwidth Internet with fingering. The PCRF may be avoided if the PGW has the "Shallow Packet Inspection" allowing us to trigger dedicated bearers when a voice RTP medium is detected. The multicast equipment in white can be added.

References

This gives a partial 2017 list of 3GPP Release 14 standards applicable to the interfaces of Figure 6.4 for the multicast service, starting with general references and then from the application server to the low-level RRC protocol.

[6.1] TS 23.246 V14.1.0 (2017-05), "Universal Mobile Telecommunications System (UMTS); LTE; Multimedia Broadcast/Multicast Service (MBMS); Architecture and functional description," Release 14.

[6.2] D.Camara et N.Nikaein editors, "Wireless Public Safety Networks 2," ISTE ed., 2016, *See chapter 9 on LTE Broadcast for Public Safety.*

[6.3] R. Liebhart, D. Chandramouli, C. Wong, J. Merkel, "LTE for Public Safety," Wiley ed., 2015, *See chapter 4 on "Proximity Services (ProSe) and 5 on Group Communication over LTE.*

[6.4] 3GPP TS 29.468 v14.2.0 (2017-07), "Universal Mobile Telecommunications System (UMTS); LTE; Group Communication System Enablers for LTE (GCSE_LTE); **MB2** reference point; Stage 3." *This Release 14 standard for AS MCPTT over LTE support "push file and streaming for the multicast coming from MCPTT". >= Release 13 supports "push through" ciphered on the MB2 interface. Interface between the CGS AS and the BM-SC.*

[6.5] TS 29.061 V14.3.0 (2017-04), "Digital cellular telecommunications system (Phase 2+) (GSM); Universal Mobile Telecommunications System (UMTS); LTE; Interworking between the Public Land Mobile Network (PLMN) supporting packet based services and Packet Data Networks (PDN)," Release 14. *The **SGmb**/Diameter interface between the BM-SC and the MBMS GW.*

[6.6] TS 29.274 V14.4.0 (2017-07), "Universal Mobile Telecommunications System (UMTS); LTE; 3GPP Evolved Packet System (EPS); Evolved General Packet Radio Service (GPRS) Tunnelling Protocol for Control plane (**GTPv2**-C); Stage 3," Release 14, *Interface GTP v2 between the MBMS GW and the MME.*

[6.7] 3GPP TS 36.344 V14.1.0 (2017-08), "LTE; Evolved Universal Terrestrial Radio Access Network (E-UTRAN); **M3** Application Protocol (M3AP)," Release 14, *Protocol M3AP ASN1 based between the MCE and the MME.*

[6.8] 3GPP 36.300 V14.4.0 (2017-10), "LTE; Evolved Universal Terrestrial Radio Access (E-UTRA) and Evolved Universal Terrestrial

Radio Access Network (E-UTRAN); Overall description; Stage 2," *Distributed MCE architecture.*

[6.9] 3GPP TS 36.443 v14.0.0 (2017-04), "LTE Evolved Universal Terrestrial Radio Access Network (E-UTRAN); **M2** Application Protocol (M2AP)," Release 14, *Multicast M2AP ASN1 based protocol between eNodeB and MCE.*

[6.10] 3GPP TS 36.331 v13.5.0 (2017-4), "LTE; Evolved Universal Terrestrial Radio Access (E-UTRA); Radio Resource Control (RRC); Protocol specification," **RRC** *Management and use of the MCCH by the eNodeB and the UE for the MBMS service. Maximum of 6 MBMS channels per eNodeB.*

[6.11] 3GPP TS 36.345 V14.0.0 (2017-04), "LTE; Evolved Universal Terrestrial Radio Access Network (E-UTRAN); **M1** data transport," *Protocol M1 between the MBMS GW and the eNodeB for the multicast service. The eNodeBs will send the synchronized time stamped data in multicast MBCH channels.*

[6.12] 3GPP TS 23.003 V14.5.0 (2017-10), "Digital cellular telecommunications system (Phase 2+) (GSM); Universal Mobile Telecommunications System (UMTS); Numbering, addressing and identification," Release 14. *For developers, the detailed coding of the protocols parameters in this chapter, SAI, etc.*

7

Integration of IMS and VoLTE in the PMR Networks and the MNOs, Details on the PCC Processing, and Access Using a Non-trusted WLAN (WiFi with an ePDG)

7.1 WiFi and VoLTE4G access to a PMR central core network

In PMR tactical networks of Chapter 2, it is usual to consider very simplified 4G EPC core netwoks. They have no SMS, no 3G access, and only the MCPTT as voice service. To provide a service continuity to the security forces, the simplest solution is to have a rather classical central EPC core network with these additions and also the WiFi access so that the PMR services can be used in police and firemen stations.

To have voice and classical PtoP SMS services under a non-trusted WLAN coverage, the most standard solution is to use VoLTE and hence it is explained in this chapter.

The prioritizing of the subscribers in a resource-limited PMR network is crucial; we explain in this chapter the PCC performed by the PCRF and PCEF (in the PGW) of the tactical or central EPCs, including the creation of dedicated bearers for the voice services by the IMS or aplication servers.

As a background, we give the call flow in Figure 7.1 for the attachment procedure in a 4G network, a tactical PMR network, or a public network. There is another access method for PMR users, a WiFi non-trusted WLAN using an ePDG as explained in Section 7.11, where you can compare with the WiFi untrusted WLAN attach procedure. The ePDG and the AAA server have the same role as an MME or SGSN in 3G.

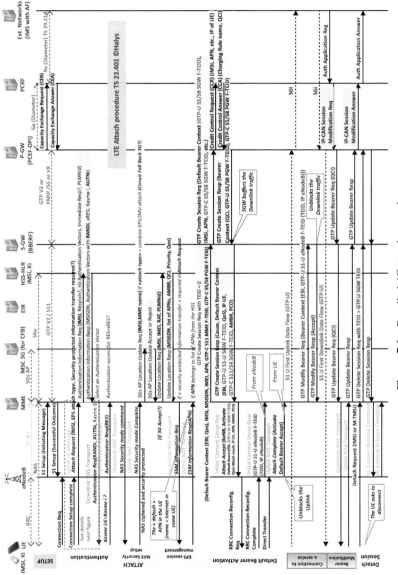

Figure 7.1 Detailed 4G attach procedure.

7.2 Operational need for VoLTE in PMR networks

Up to now in this book, the focus has been on intervention type public safety forces with a telecommunication service along lines not far from that used in the land armies. But most of them have more usual daily tasks which require a much broader coverage than just push-to-talk. They must also be able to call or be called with any number. VoLTE becomes an evidence because the simplicity and performance of the infrastructure which will soon be generalized will entail cost reductions, and this is a major concern for the security forces moving from proprietary technologies and networks to LTE. Also VoLTE terminals do not need the provisioning of a SIP client which choice by the customer is broad with rather frequent defects and for some recent models, depend of a proxy notify server to be waken up when they idle to save power.

An IMS core is needed for VoLTE, so it is explained below in Section 7.4.

7.3 Reminder of the VoTT architecture for a pure VoIP MNO

7.3.1 Public identity for VoTT VoIP vs LTE

The "public identity" for VoTT VoIP *is the MSISDN* and is configured in the SIP phone. It does not need any provisioning for the VoLTE UE; the public identity is the IMSI from the SIM card. The "public identity" is provided in the SIP REGISTER sip: public-identity@domain-name. If there is an AKA authentication, the MSISDN is included in the response to the SIP 401 sent by the S-PCSF. A *core network working both* with pure VoIP SIP phones and UE LTEs must have an HLR-HSS which provides the user data from this public identity by trying this *as an IMSI first and* then if not found *as an MSISDN*. Such an HLR-HSS is assumed in the sequel.

7.3.2 VoTT VoIP network architecture

A pure VoIP MNO has only MMEs in its network but the subscribers can switch to 3G when they roam in a foreign partner. The diagram of the core network must be like Figure 7.2. The TAS is a classical VoIP server without interfaces to the HLR-HSS as explained in [0.10, Chapter 15].

Figure 7.3 adds front-end roaming hubs to allow multi-IMSI roaming, either with "direct IMSI sponsors" (bottom left) or with "IMSI aggregators" such as Telna (USA) which have several direct "IMSI sponsors." As a result, there can be up to 10 IMSIs in the SIM cards. There is a PCRF but which

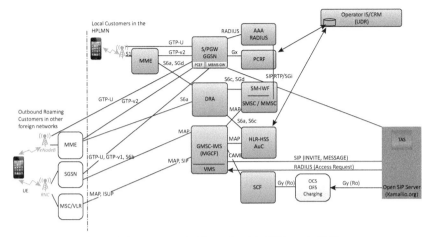

Figure 7.2   Pure local 4G MNO with OTT mode VoIP.

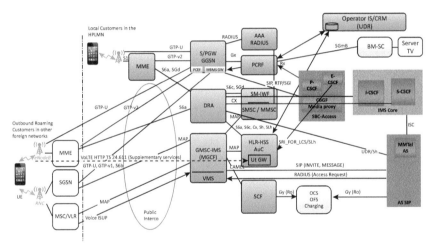

Figure 7.3   Pure local 4G MNO with ISUP connectivity and an IMS core.

provides only the QoS customization depending on the subscription. There is no session border controller (SBC) (P-CSCF) capable of analyzing the Session Data Protocol (SDP) with the media and triggering the PCRF for the creation of "dedicated bearers" with the PCEF for improved VoIP QoS with guaranteed bit rate (GBR). This is provided by a core IMS VoLTE introduced in the next section.

7.4 IMS-based PMR network architecture for the services

Chapter 15 of [0.10] has explained VoIP OTT-based PMR for voice and SMS services which is quite comprehensive except some services which are available in 3G. Terminals with the VoLTE client have the "Supplementary Services" (call barring, call forwarding) thanks to IMS.

7.4.1 Equivalence between 3G/2G notions, VoLTE/IMS, WiFi EAPsiim/VoTT, and SIP VoTT

Table 7.1 shows that WiFi EAPsim is close to VoLTE; it is the base of WiFi access (WLAN) to 4G services.

The SBC contains the P-CSCF, which uses Rx/Diameter to communicate with the PCRF by AAR, and the RTP-Proxy, which routes the RTP traffic through three possible interfaces:

- SGi or Gi of S/PGW-GGSN for GSM or LTE data
- Internet with wire or WiFi interfaces
- the GMSC/VLR/SSF-IWF-IMS interfacing the PSTN and the SIP.

The following IMS logical entities intervene:

- P-CSCF "Proxy-CSCF" analyzes the SIP protocols in the IP messages and decides the S-CSCF,
- S-CSCF "Serving PCRF" modifies the HSS pour connaître to set the S-CSCF with a SAP. This is the "presence server" of the SIP service which performs the routing of the outgoing calls, in particular the GMSC/VLR/SSF-IWF (MGCF) if the destination of a call or an SMS is not SIP registered. It also receives from the I-CSCF the calls coming from the GMSC/VLR/SSF-IWF-IMS (MGCF).
- I-CSCF "Interrogating CSCF" which interrogates the HSS with a LIR (location information request with the same role as SEND_ROUTING_INFO MAP) to find the S-CSCF and addressing the INVITE to this S-CSCF (see Figure 7.4 of an incoming call).

7.4.2 Equivalence between 3G/4G notions and the equivalent in IMS (mobility management of Cx/diameter)

There is no necessity for a separate authentication with quadruplets (4G) or quintuplets (3G), as there was already an authentication with SAI and then a UL to register the IMSI in the HLR-HSS if the access is from a mobile network and not through the Internet. IMS is a user plane application with

Table 7.1 Equivalence between ISUP and SIP

3G–2G notions (circuit voice services)	Equivalent VoLTE 4G/IMS	WiFi EAPsim [0.9, Chapter 8]	SIP VoTT
PLMNid	Visited Network Identifier		
IMSI	Public Identity in the VoLTE case example: sip:262011234567890@ims. mnc001.mcc262.3gppnetwork.org In the URI parameter of SIP REGISTER INVITE sip:262011234567890@ims. mnc001.mcc262.3gppnetowork.org	In the URI parameter of SIP REGISTER INVITE sip:262011234 567890@ims. halys.fr	Provided in the response to SEND_IMSI (MAP sent by the GMSC/VLR/ SSF-IWF-IMS to the HLR)
MSISDN	• Public Identity for VOTT VoIP tel: msisdn example: tel: +491234567890 • Authentication in the 2nd SIP REGISTER in response to a SIP 401 of the AKA authentication provided in SAA/ Cx	Provided in the response to the RESTORE DATA sent by the GMSC/VLR/SSF-IWF-IMS	In the URI parameter of the SIP REGISTER INVITE sip:33970675 065@ims. halys.fr
IMSI	User-Name example: 262011234567890@ims.mnc001. mcc262.3gppnetwork.org		
Visited MSC/VLR GT	Server Name (S-CSCF): example: sip:scscf.ims.mnc001.mcc262. 3gppnetwork.org:5060		
HLR → MSC/VLR Parameters of INSERT SUB-SCRIBER DATA Req (MSISDN)	HSS → S-CSCF Parameters of Server Assignment Answer: User-Data, User-Profile <IMSSubscription> (MSISDN)		
ISUP IAM	SIP INVITE		
—	SIP 100 Trying		
ISUP ACM	SIP 180 Ringing		
ISUP ANM	SIP 200 OK		

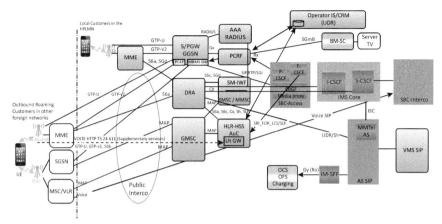

Figure 7.4 Pure local 4G MNO with SIP connectivity and a full IMS.

its own Diameter protocols. If the access is from a mobile network, the simplified sequence could be:

- Authentication (SEND AUTHENTICATION INFO) with a RAND random "challenge" between the SIM and the AuC with separate RES response with the common secret key Ki.
- Registration of the localization (UPDATE LOCATION) MME or SGSN → Server-Name IMS.
- IMS Authorization and then registration of the user's S-PCSF

while with a fixed or WiFi Internet access to IMS:

- Authentication with SIP-Digest [7.12] or EAP/SIM with a RADIUS server
- IMS authorization and then registration as above.

7.4.3 Incoming call (protocol Cx/diameter)

See Figure 7.2, the Table 7.2 below shows the Cx/Diameter messages used for incoming calls and their MAP equivalent.

7.4.4 IMS subscriber's services' management (protocol Sh/diameter)

The requests are from the TAS to the HSS (Table 7.3).

The core network with IMS is given in Figure 7.3:

The operator's information system (customer relation management (CRM)) has a graphic interface to manage the customer's profile and import the SIM cards; it is the 3GPP user data register (UDR). In the implementation

Table 7.2   Cx messages for incoming calls

Notion 3G/4G (MAP TS 29.002)	Equivalent IMS (Cx TS 29.329)
GMSC → HLR incoming call **SEND ROUTING INFO** Req (IMSI)	I-CSCF → HSS incoming call Location Information Request (**LIR**) (Public identity)
In VOTT [0.10, Chapter 15] returns the GT of GMSC/VLR/SSF "visited" by a SIM registered subscriber	Allows to know the S-CSCF "visited" by a subscriber
SEND ROUTING INFO Cnf (VLR GT, Roaming Number)	Location Information Ack (**LIA**) (S-CSCF Identity)

Table 7.3   Sh messages used in the IMS procedures

Notion 3G/4G (MAP TS 29.002)	Equivalent IMS (Sh TS 29.239) [11.15]
ANY TIME SUBSCRIPTION INTERROGATION (interrogation of HLR-HSS by the SCF of the prepaid subscriber from the GT in its subscriber's profile.	TAS → HSS User Data Request (**UDR**)
	User Data Ack (UDA)
ANY TIME MODIFICATION (update of the subscriber's profile in the HLR-HSS by the SCF)	TAS → HSS Profile Update Request (**PUR**)
	Profile Update Ack (PUA)
No equivalent	TAS → HSS Subscribe Notifications Request (**SNR**)
	Subscribe Notifications Ack (SNA)
NOTE SUBSCRIBER DATA MODIFIED (the HLR-HSS informs the SCF that the subscriber's profile has been updated	HSS → TAS Push Notifications Request (**PNR**) (User identity, user data)
	Push Notifications Ack (PNA)

above, HLR-HSS and PCRF *are eventually from the same supplier*; UDR is a common database for HLR-HSS and PCRF which handles the QoS depending on the subscriber's profile, the requested APN, and the usage of the different cells (dynamic optimization, see patent [14.3] in |0.9]). In some variants the HLR-HSS profile database is used by the PCRF.

In a more "from the book implementation," HLR-HSS and PCRF have their own databases fed by the LDAP/Ud protocol from the UDR.

The IMS core is colored orange with I-CSCF et S-CSCF ("presence server" in VoTT), the P-CSCF(SBC) is in pale red, and the TAS ("MMTel") is in violet and has the Sh interface with the le HLR-HSS.

We have included the BM-SC of the multicast server used by a "TV Server" but it is not an IMS-related function; it is for the PMR "push-to-talk" AS.

The Cx/Diameter and Dx/Diameter interfaces use the same protocol [7.11]. Cx is I-CSCF→HSS and Dx is I-CSCF<→SLF (the server location function which finds the HSS name when it is not known from a previous registration).

7.5 Call flow of the IMS services

Compared with Figure 7.2, the operator connectivity uses SIP with the SBC-Interco. The GMSC handles only MAP, as the "media conversion MGCF" is now handled by the core IMS. The VMS has a SIP connectivity (all SIP → SIP, SIP → PSTN, PSTN → SIP calls are then handled).

The routing of calls to VMS is handled by the AS SIP, which whenever a subscriber registers interrogates the HLR-HSS with a UDR/Sh/Diameter to get the subscriber profile. The result also contains the triggers for prepaid charging handled by an IM-SSG which has the Gy/Ro interface with the OCS.

7.5.1 IMS registration: Voice calls

The registration procedure is a major architecture feature so that mobile terminals may receive calls and SMS whichever network they are using. All reception of calls and SMS must be anchored to their HLR-HSS register using the public identity. This section and Figure 7.4 describes the registration process in an IMS network which allows the reception of incoming calls. In Chapter 3, the registration process is explained to ensure the reception of SMS-MT.

Figure 7.5 is the UE registration in the IMS of the HPLMN or when it is VoLTE roaming in a partner's network using "IMS Home Routed Roaming Architecture" [7.13, Figure 5]. The outgoing calls, even to local numbers, while roaming trombone through the IMS in the HPLMN through the GRX (IPX) which may degrade the quality with latency, but this is the current implementation of VoLTE roaming services.

The initial "SIP Register" is sent to the IPv4 or IPv6 address or domain name of the P-CSCF contained in the EFp-CSCF file of the SIM card [7.7]. The URI has a public identity based on the IMSI of the Efipmi file (not the MSISDN as in VoTT) where the URI is derived from the

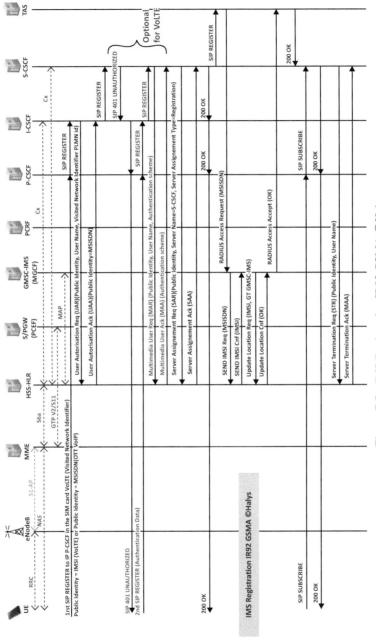

Figure 7.5   IMS Registration IR92 GSMA ©Halys.

User-Name = MSISDN and the domain name manually configured in the SIP client. In VoLTE, the iSIM card ("i" for iSIM) has all the configuration and there is no provisioning required by the customer.

Let "SIP register" also contain the parameter "Visited-Network" (see an example):

P-Visited-Network-ID: "mnc001.mcc262.3gppnetwork.org"

The P-CSCF proxy then sends the "SIP Register" to the I-CSCF, which sends to the S-CSCF.

The I-CSCF uses the Visited-Network to interrogate the HLR-HSS with UAR containing "Visited-Network" to check if the UE is authorized to register in this VPLMN. If yes, it sends a SAR (equivalent of an UPDATE LOCATION) to the HLR-HSS while supplying the S-CSCF Server-Name (which is the equivalent of MSC/VLR GT).

7.5.1.1 Authentication of the subscriber, VoLTE and OTT VoIP compatible core IMS: MAR and MAA/Cx messages

It is desirable that the same core IMS and core EPC can simultaneously handle VoLTE and OTT VoIP clients which use two different authentication schemes. The User-Access-Request UAR gets an answer UAR which contains the AVP Supported-Features and indicates to the S-PCSF which is the case for a particular MSISDN. Then the P-CSCF sends a MAR with the selected authentication scheme.

Then the P-CSCF sends a "SIP 401 unauthorized" answer to the first SIP INVITE which has SIP-Authenticate as parameter. The mobile will compute its own answer to the RAND challenge (also called NONCE) and send a second SIP INVITE with the answer in the "Authorization" parameter (equivalent of SRES in 3G or 4G computed by the SIM). The P-CSCF compares "Authorization" in the SIP INVITE and SIP-Authorization in the MAA from the HSS, and if not equal will reject.

The HLR-HSS knows whether it should use AKA 3G with milenage (the Public-Identity *has 15 digits*) or legacy SIP Digest [2.12] (the Public-Identity *has less than 15 digits*) without any profile setup and can then *mix VoLTE capable customers and SIP VoTT customers.*

A VoLTE client using 4G or 3G is already authenticated, and this step is redundant; this is why it is in red and optional for VoLTE in Figure 7.5. So, the P-CSCF if it sees a Public-Identity of 15 digits could easily skip the MAR.

Table 7.4   Content of the MAA AVPs for a VoLTE and OTT VoIP compatible HSS

AVP Contents in the MAR (Multimedia-Auth-Request) and MAA	VoLTE (Milenage)	OTT VoIP (fixed or mobile)	OTT VoIP proprietary
Public-Indentity	IMSI (read by the "supplicant" layer in the UE) *208940000001065@ mnc094.mcc208. 3gppnetwork.org*	MSISDN set in the SIP client *33970675065@ims. halys.fr*	MSISDN set in the SIP client
Private-Identity	MSISDN (provided by the HSS)	MSISDN (set by the HSS) identical to the Public-Identity	MSISDN (set by the HSS) identical to the Public-Identity
SIP-Number-Auths-Items in MAR	N = number of 3G vectors	1	1
Authentication-Scheme in MAR and MAA	AKAv1-MD5 (3GPP AKA)	SIP Digest	Cirpack Fix
SIP-Item-Number in MAA	i starting at 1 for the 1rst vector	1	1
SIP-Authenticate in MAA	RANDi = random 16 bytes AUTNi from Milenage 16 bytes concatenated	NONCE = random 16 bytes	NONCE = random 16 bytes
SIP-Authorization in MAA	XRESi	Response = MD5(MD5(A1): NONCE:MD5(A2))	User password
Confidentiality-Key	CKi derived from Milenage	random 16 bytes	random 16 bytes
Integrity-Key	IKi derived from Milenage	random 16 bytes	random 16 bytes

Trace of the Multimedia Autn Answer from the HSS to the S-CSF

This is the equivalent of a SAI and the most important Cx message to understand how the compatibility VoIP OTT and VoLTE is implemented. Here the Public-Identity is MSISDN, and hence the SIP-Digest authentication is performed, the AVP SIP-Digest-Authenticate is filled, and there is a single vector.

Send Cx (TS 29.229 Rel.14)

```
Result: DIAMETER_SUCCESS

AVPs:
AVP_Session_Id (263) l=63, f=-M-
val=scscf1.epc.mnc094.mcc208.3gppnetwork.org;1495889334;281
AVP_Vendor_Specific_Application_Id (260) l=32, f=-M-
    AVP_Vendor_Id (266) l=12, f=-M- val=10415
    AVP_Auth_Application_Id (258) l=12, f=-M- val=16777216
AVP_Auth_Session_State (277) l=12, f=-M- val=NO_STATE_MAINTAINED (1)
AVP_Origin_Host (264) l=48, f=-M- val=hlrhss.epc.mnc094.mcc208.
3gppnetwork.org
AVP_Origin_Realm (296) l=41, f=-M- val=epc.mnc094.mcc208.3gppnetwork.
org
AVP_Result_Code (268) l=12, f=-M- val=2001
AVP_User_Name (1) l=23, f=-M- val=208940000001065
AVP_Public_Identity (601) l=23, f=VM- vnd=VND_3GPP val=33970675065
AVP_SIP_Number_Auth_Items (607) l=16, f=VM- vnd=VND_3GPP val=1
AVP_SIP_Auth_Data_Item (612) l=240, f=VM- vnd=VND_3GPP
    AVP_SIP_Item_Number (613) l=16, f=VM- vnd=VND_3GPP val=1
    AVP_SIP_Authentication_Scheme (608) l=28, f=VM- vnd=VND_3GPP
val=Digest-AKAv1-MD5
    AVP_SIP_Authenticate (609) l=28, f=VM- vnd=VND_3GPP
val=723669684F2A23232E333B6C4D606000
    AVP_SIP_Authorization (610) l=18, f=VM- vnd=VND_3GPP
val=6D617263656C password 'marcel'
    AVP_Confidentiality_Key (625) l=28, f=VM- vnd=VND_3GPP
val=459453B36757B20663A59D735379AE35
    AVP_Integrity_Key (626) l=28, f=VM- vnd=VND_3GPP
val=52B92DA12983AE496C8121E03911B9E9
    AVP_SIP_Digest_Authenticate (635) l=80, f=V-- vnd=VND_3GPP
        AVP_Digest_Realm (104) l=20, f=-M- val=ims.halys.fr
        AVP_Digest_QoP (110) l=8, f=-M- val= auth
        AVP_Digest_HA1 (121) l=40, f=-M-
val=5C2C5CCD0F39B4F6A5FDCB3D921B793F HA1 computed by HSS
```

7.5.1.2 Registration in the HSS to be able to receive calls and SMS

Legacy mode with RADIUS

Some manufacturers have a possibility to receive calls to SIP phones with a legacy RADIUS SIP server as in Figure 7.2. The TAS does not have Diameter and performs a RADIUS Access-Request to the GMSC-IMS which is the MGCF function, and acts as a proxy to the HLR to send and UPDATE-LOCATION. It allows the HLR to respond the GT of the GMSC-IMS when interrogated by a SEND-ROUTING-INFO, coming from the GMSC-IMS, or a SEND-ROUTING-INFO-FOR-SM in the incoming call or incoming SMS procedure.

IMS mode (the same for VoLTE or legacy OTT VoIP)

The P-CSCF sends a Server-Assignment-Request (SAR) containing REGIS-TRATION and the Server-Name = S-CSCF, which is the equivalent of an UPDATE-LOCATION with the VLR or the MME. For incoming calls, the HSS will be able to respond with this Server-Name when interrogated by the LIR of an incoming call, as illustrated by Figure 7.4.

7.5.1.3 De-registration of a subscriber

This happens when the SIP application is stopped. The S-CSCF will send a Server-Assignment-Request (USER DEREGISTRATION) to the HSS. This is the equivalent of a CANCEL LOCATION.

7.5.2 Handling of incoming calls or SMS from the PSTSN or the SS7 network

7.5.2.1 Emergency call handling in IMS with calling party localization

The TAS needs the calling party localization to route the call to the closest local emergency center 112. We assume a simple system where the local-ization is indicated by the postal code of the visited cell (ARCEP regulation in France). It is provided by the visited station UTRAN or eUTRAN from its setup. More precise localization could be provided (see [0.10, Chapters 2 and 12]).

As explained in the 3GPP spec Rx ([11.18], Annex 10 of [0.10]), when the IMS P-CSCF sees an INVITE to 112, it triggers an AAR in Figure 7.4 to ask the latest known localization by the PCRF by setting the parameters:

Specific-Action AVP   ACCESS_NETWORK_INFO_REPORT
Required_Access_Info AVP   USER_LOCATION ou MS_TIMEZONE

Remember that the PCRF keeps track of the subscriber's location from the messages with the PGW if the parameter "Change Reporting Action" has been set in the create session response of the PGW to the MME [7.9], and then the MME informs the PGW of all the changes. The PGW informs the PCRF with Credit-Control-Request (CCR)/Gx. All this signaling is not represented but essential to understand how it works.

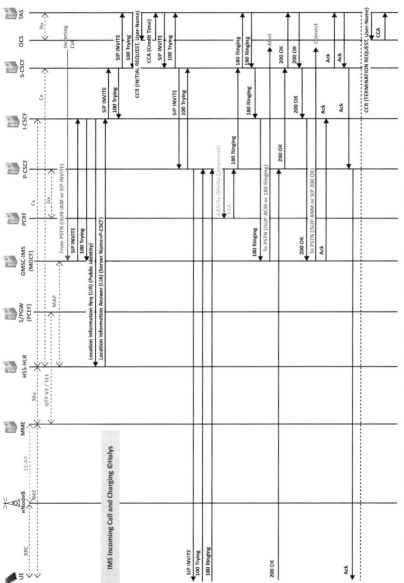

Figure 7.6 PSTN incoming call with localization of the calling and called party then IMS outgoing call.

This allows the PCRF to answer at any time to the AAR from the TAS by sending a RAR to the P-CSCF which contains:

Specific Action AVP   ACTION_NETWORK_INFO_REPORT
User-Location-Info AVP CGI, SAI, RAI, TAI, ECGI (la "cellule") ou LAI
SGSN-MCC-MNC   Visited Network Identifier.

The RAR and RAA messages for emergency calls are orange colored. All this assumes that the TAS has the cell map; otherwise, it knows only the visited network identifier and also that the entire chain MME-S/PGW-PCRF complies entirely with 3GPP (the implementation of change reporting action [2.9] is essential; otherwise, there is no use that that the P-CSCF asks the localization to the PCRF).

When it receives the User-Location-Info in the RAR, the P-CSCF inserts it in the SIP INVITE sent to the TAS according to [7.10] that is an additional P-Header:

P-Access-Network-Info = "P-Access-Network-Info" HCOLON

access-net-spec *(COMMA access-net-spec)

access-net-spec = (access-type/access-class)

*(SEMI access-info)

and access-info contains depending on the case CGI, SAI, etc. The TAS has a table which for all cells gives the closest emergency center to call.

7.5.2.2 SMS

The call flow is identical: instead of an incoming ISUP IAM or SIP INVITE, we have MAP FORWARD SM MT or SIP MESSAGE. In Figure 7.4, replace INVITE by MESSAGE and the messages 100 Trying and 180 Ringing do not exist; there is directly a 200 OK when the SMS is received by the SIP client.

7.5.2.3 Charging of the calls and SMS

Figure 7.4 shows the charging of the outgoing call to the roaming UE with the Ro/Diameter with CCR messages. The LIA gives the S-PCSF address but the I-CSCF knew the Visited-Network-Identifier from the Registration IMS of Figure 7.3 and provides it in the INVITE to the TAS. If the UE is roaming, the roaming call rate will apply. Based on the IMS subscription, the TAS knows if a prepaid customer is concerned and triggers a charging by Ro/Diameter with the OCS.

The data use charging is done by the PGW using Gy/Diameter or sometimes using RADIUS [7.8] using quotas and thresholds so that the PGW asks a new quota when it is closed to completion. The call flow for Gy and RADIUS is similar. In RADIUS, the real-time Access-Request may be used while Accounting-Request may not be. The PCC must be capable of measuring the data volume of a session. In [0.9, Chapter 4], another RADIUS data charging system is explained which uses only regular Accounting-Request (Interim-Interval) to inform about the usage until the RADIUS server sends a Disconnect, but [7.8] seems to be mostly used.

1) Radius data charging [7.8]: Subscriber opens a data session

```
      GGSN-S/PGW (PPC)  . . . . . . .Server(AAA) Prepaid System(PPS)
  | Create Session(GTPv2)|
  |  Calling-Station-ID  |
  |  Called-Station-ID   |
    | ----------------->| Access-Request                          |
  |                     | User-Name: 0000                         |
  |                     | User-Password: marcel                   |
  |                     | Called-Station-ID= APN                  |
  |                     | Calling-Station-ID: MSISDN              |
  | Quota request       | Service-Type:Authenticate-only          |
  | PGW gives its       | Prepaid-acct-Capability:Volume,
                        | Duration                                |
  | capabilities        | Acct-Session-Id: 00336503               |
  |                     | ------------------------------------->|
  |                     | Access-Accept                           |
  |                     | User-Name: 0000                         |
  |                     | Service-Type:Authenticate-only          |
  | PPS sens an initial | Prepaid-Acct-quota(QID,Volume=5Mb,
                        | Threshold=4Mb)|
  |   5Mb quota         |    ---------------|  Credit resevation
```

Then the PGW must send an Accounting-Request (Start) which should not get an Accounting-Response according to [7.8] while this not completely standard real example has one, it is used to inform the AAA (PPS) that the data session has started. Otherwise, the PPS would know it only when it receives the first quota refresh demand; after some time, the PPS sends a RADIUS Disconnect to the PGW.

```
  | ------------------->| Accounting-Request                      |
  |                     | User-Name: 0000                         |
  |                     | Framed-IP-Address: 10.230.138.8   |
  |                     | Called-Station-ID= APN                  |
  |                     | Calling-Station-ID: MSISDN              |
  |                     | Acct-Status-Type:Start                  |
  |                     | Acct-Session-Id: 00336503         |
  |                     | etc.. autres AVP
                          Vendor-Specific                        |
  |                     | ------------------------------->|
  |                     | Accounting-Response                     |
  |                     | User-Name: 0000                         |
  |                     | <-------------------------------|
```

2) When the 4 Mb threshold is reached, the PGW asks a new quota to the PPS; it provides what has been used so far and asks the permission to pursue.

```
|                          | Access-Request                         |
|                          |  User-Name: 0000                       |
|                          | Called-Station-ID= APN                 |
|                          | Calling-Station-ID: MSISDN             |
| Athorisation demand      | Service-Type:Authenticate-only         |
|  Used Volume so far      | Prepaid-Acct-quota(QID,Vol used=4Mb,   |
|                          |   reason=(3)threshold reached)         |
|                          | Acct-Session-Id: 00336503              |
|                          | -------------------------------------->|
|                          | Access-Accept                          |
|                          | Service-Type:Authenticate-only         |
|                          | User-Name: 0000                        |
| PPS sends a 4Mb          | Prepaid-Acct-quota(QID,Volume=4Mb,     |
|                          |   Threshold=4Mb)                       |
|   additional quota       | <--------------------------------------|
```

3) The subscriber disconnects the data session [7.8, page 70]

```
|  Delete Session(GTPv2)|
|  Calling-Station-ID   |
|  Called-Station-ID    |
| -------------------->| Access-Request                          |
|                      |  User-Name: 0000                        |
|                      |  Called-Station-ID= APN                 |
|                      |   Calling-Station-ID: MSISDN            |
|                      |   Service-Type:Authorise-only|          |
| Used Volume so far   | Prepaid-Acct-quota(QID,Vol used=2Mb,    |
|                      |   reason=(8)serv terminated) |          |
|                      | Acct-Session-Id: 00336503               |
|                      | ------------------------------------->  |
|                      | Access-Accept                           |
|                      | Service-Type:Authenticate-only          |
|                      | User-Name: 0000                         |
| PPS sends            | Prepaid-Acct-quota(QID,Volume=3Mb)      |
| the reimbursed volume| --------------------------------------|
Reimburse the credit 8.7: RADIUS[8] charging of the prepaid data
service
```

7.5.3 Outgoing SMS or voice calls to the SS7 network or the PSTN

7.5.3.1 SMS

The SMS service must not be a simple "chat" but benefits from the usual retry mechanism if the destination is not reachable. For all destinations, whether they are SIP registered or not, the TAS must send the SMS to the GMSC-IMS which translated and send it to the SMSC as if it was coming from

3G or 4G. The SMSC will handle the sending with the usual retry and ALERT scheme. If the destination is SIP registered, it is translated MAP→SIP by the GMSC-IMS.

7.5.3.2 Charging
Figure 7.4 shows the charging for incoming calls to a roaming subscriber. For outgoing calls and SMS, it is identical and uses the Ro/Gy interface of the OCS.

7.5.3.3 3G voice calls to a subscriber in a mobile network: Non-IMS case
In this case, the SRI/MAP if the call comes from the PSTN returns an MSRN. If the call is from the SIP, the LIR/Cx returns WilcartedPrivateIdentity = MSRN. In both cases, the GMSC will call this MSRN over the PSTN.

Often the subscriber is visiting the MSC/VLR of a partner network national or foreign partner and a solution should exist to avoid the "tromboning" arising from the three conditional call forwarding (busy, no response, and not reachable).

7.5.4 Anti-tromboning of the calls to mobiles 3G: The SORTA method [7.16] passive camel monitoring and MAP call transfer package

7.5.4.1 The general tromboning for outgoing calls to 3G with conditional forwarding of unsuccessful calls
Anti-tromboning is commercially important when the VMS service is provided as more that 30% of the calls fail (unreachable, no reply, and busy) and when forwarded back to the VMS are charged by the VPLMN. The tromboning situation of Figure 7.7 arises:

- when a call is received by the telephony gateway GMSC either from the PSTN ISUP 1) bottom left, or from the SIP 1) bottom right. The SIP call may come from the TAS SIP or from the Border SBC with a SIP connection to the PSTN.
- the destination is a mobile registered in a 2G or 3G VMPLM and the mobile's profile includes conditional call forwarding (no response, busy, and not reachable).

The international telephone network (notably ISUP) does not guarantee that the parameter Call Reference [2.17] is relayed (optional parameter) and only

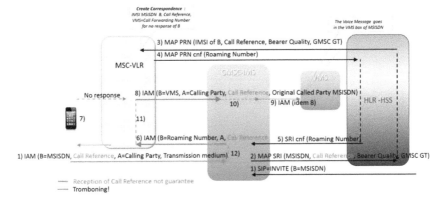

Figure 7.7 Tromboning case: Call forwarding to VMS for lack of response of the call.

the Header of the IAM plus the calling party number and original called number can be trusted. This is the reason why a general anti-tromboning is not trivial as a matching between the outgoing call 6) and the tromboned call 8) is required. The call reference if available provides this matching which is an absolute requirement for the 3GPP anti-tromboning of Section 2.3.4.2 *(needs the call reference* in the IAM 6) of Figure 7.7).

Otherwise, one has to use one of the alternative anti-tromboning methods:

- 1) Incoming call from the PSTN or from the SIP server or the Border SBC is received by the GMSC-IMS. The presence of a call reference in the ISUP IAM is not guaranteed (for example, Orange FR ISUP interco) but it is unimportant.
- 2) The GMSC-IMS sends a SEND ROUTING INFO Request to the HLR with the destination MSISN, the received Call reference if any or one it creates, and the bearer quality GSM computed from the medium requirement contained in the call ISUP or SIP INVITE. This is to obtain the MSRSN.
- 3) The HLR-HSS sends a PROVIDE ROAMING NUMBER V3 with the IMSI and this call reference (and the bearer quality for visio calls H324-M) as well as the GMSC-IMS GT. The VLR returns 4) the MSRN roaming number, which is included in the SEND ROUTING INFO Cnf 5).
- 6) The GMSC-IMS makes an outgoing call 6) to the VLR using the MSRN as called number and including the call reference (which it has received or generated).

- 7) and 8) the UE rings, no reply. The VLR uses the profile Cfnoreply of B to call 8) the ForwardedToNumber (the VMS). The ISUP IAM 8) *may* include the call reference *if it was relayed by the network, from the GMSC-IMS to the VLR.*
- 9) The GMSC-IMS will transmit the IAM to the VMS including the call reference *if it was relayed by the network from the VLR to the GMSC-IMS.*
- 10)–11)–12) The VMS answers with ACM and ANM relayed back to the original call 1), all the call legs being connected. There is a trombone 12)–11)–10) between the original call 1) and the VMS 9) through the backward call 8).

The VPLMN charges the leg 8) to the HPLMN. If VPLMN = China, this is more than 2 euros/min. The GMSC returns a MAP RESUME CALL HANDLING confirmation to ANTI-TROMBONING indicating that the call transfer to VMS has been successful.

Then the VMSC (Lebanon) sends a CAMEL RELEASE (8) to the SCP (Armenia) which ANTI-TROMBONING monitors and shows that the tromboning attempt has been successfully canceled.

All the anti-tromboning methods below will work *for any ForwardedTo numbers as* they are implemented by the HPLMN GMSC and its SCP in the SORTA case, not by the destination equipment such as a VMS. The SORTA method of 2007 is thus improved *by being implemented not in a VMS but more generally by an SCP with a GMSC-IMS* handling the standard MAP Call Transfer Package.

7.5.4.2 Anti-tromboning #1, 3GPP MAP method, not applicable in practice

If a) there was an anti-tromboning agreement and the proper "Optimal Routing" configuration between the VPLMN and the HLPMN (the MAP Call Transfer Package), b) the PROVIDE ROAMING NUMBER is MAP V3 (to include the Call Reference and the GMSC GT, the VLR sends a RESUME CALL HANDLING to the GMSC-IMS with the Call Reference and the ForwardedToNumber. The GMSC-IMS will cut the IAM Leg 6) and connect the Leg 1) directly to the ForwardedToNumber as in Figure 7.8.

Unfortunately, this very simple method works only when there is a cooperation. It exists if the two networks belong to the same group (Vodafone, Orange, etc.) or is imposed by a regulatory body. Anti-tromboning is mandatory in the European Union. In other cases, it is not in the interest of the

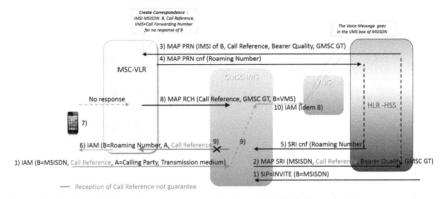

Figure 7.8   The 3GPP anti-tromboning with the MAP Call Transfer Package requires a cooperation of the VPLMN.

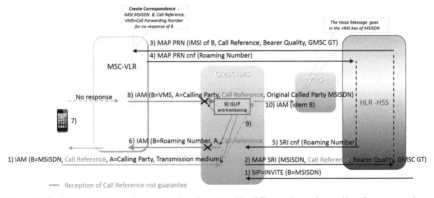

Figure 7.9   Pure ISUP anti-tromboning by the GMSC requires the call reference to be a secure method.

VMPLMN which loses the revenue of the call 8) to the HPLMN and this method *is quite rarely implemented in non-regulated cases.*

7.5.4.3 Anti-tromboning # 2, pure ISUP handling by the GMSC-IMS: simple but not quite general (Figure 7.9)

Original Call Reference present in the tromboned "Backward Call" 8)

Call reference taken from the original called number in 8)

In the absence of call reference in the IAM 8), a method could be to use the original called party as a "call reference." However, there could be two calls simultaneously sent to the roaming UE, and thus there is a risk that the wrong leg be connected to the VMS with a wrong calling party.

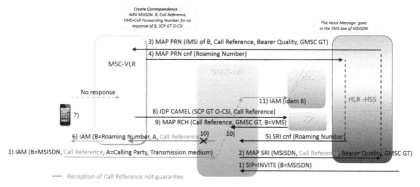

Figure 7.10   SORTA anti-tromboning with CAMEL and RESUME CALL HANDLING.

7.5.4.4 Anti-tromboning # 3, the SORTA method [7.16]: CAMEL and use of MAP RESUME CALL HANDLING by a CAMEL HANDLING SCP: rigorous (Figure 7.10)

This method is rigorous as there is no ambiguity on the call reference which is relayed in the *Camel IDP signaling* from the MAP PROVIDE ROAMING NUMBER *signaling* but the most complex. It uses a camel handling SCP and is justified because it also handles the "bill shock" and the "number correction."

- In case of no response, the call 6) must be forwarded to the VMS. As the subscriber has a prepaid profile, it is preceded by an IDP Camel interrogation 8) of the credit. *This IDP contains the call reference from the PRN signaling and is transmitted by SS7 and then is always present. This* is not the case when it is transmitted by ISUP.
- The SCP recognizes that this is a call forwarding (presence of a call diversion parameter, of an original called number) and sends a MAP RESUME CALL HANDLING to the GMSC-IMS GT with the called number (the VMS, the home, etc.) and the call reference. As in 9), the call 1) is connected to the VMS or home 11).

It works almost exactly as the 3GPP anti-tromboning except that the RCH is sent to the GMSC by the HPLMN's SCP instead of from the VLR the cooperation of which cannot be accounted for.

7.5.4.5 Trace details CAMEL and MAP of the SORTA anti-tromboning [7.16]

The number of the CAMEL outbound roamer of Armenia (+374) (roaming in Lebanon (+961)) is +37494987654; it has VMS provisioned (forward-to-number = +37493297130) and the no-reply conditional forward-ing condition at 10 s.

We make a call to this number from +33140123456 in France (+33).

8) After the call 6) has rung for 10 s, the SCP in the HPLMN receives an IDP.

```
*** InitialDP indication ***
    CallReferenceNumber=1584B40001,
     IMSI=+283052000345678,
    called Party BCD Number:    absent,    (BCD means Before Call
    Deviation)
    called Party Number= present +37493297130,   (that shows that
    the call is redirected)
    calling Party Number= +33140123456,   (calling party from
    France)
    original Called PartyID= +37494987654 (the called number of
    Armenia)
    redirection Information: no reply
    redirecting Party ID= +37494987654 (also the called number of
    Armenia)
```

It is recognized as call forwarding because the redirection information is *present* in the InitialDP as well as the original called party.

9) The SCP sends a MAP RESUME CALL HANDLING to the GMSC=+37493297100.

```
*** MAP RESUME CALL HANDLING request ***
IMSI = +283052000345678)
CallReference=1584B40001
Forwarding Options= no reply
Forwarded-to-Number = +37493297130.
```

As you see, the "redirectionInformation" = **noreply** from the InitialDP is set back in the RESUME CALL HANDLING parameter "Forwarding Options," so that the IAM ISUP message (7) of Figure 7.3 will also have it. If the VMS has different prompt messages when the called subscriber does not reply ("your correspondent has not answered" or "your correspondent is busy"), it will play the prompt corresponding to the case.

The efficiency of this method is that it works in a secured manner even if there is no ISUP transmission of the call reference.

7.6 What brings VoLTE, interest of IMS for the combined mobile-fixed service

In [7.13], there is the description of the VoLTE client; it provides the same functionalities than a 3G set for the "Supplementary Services" (call forwarding and call barring). The implementation of VoIP in an MNO is greatly simplified because there is no need to provision the SIP client (no user-name, no password, and no domain that the users must set). This is the major and main advantage compared to VoIP OTT of [0.10, Chapter 15] which besides has exactly the same quality for voice and SMS services. Some SIP clients such as Linfone use only the SIP interface for all calls including emergency calls. IMS is not a factor of voice quality; it is the same in VoLTE and VoIP OTT both in 4G. VoLTE will be rather general in 3 years from 2017 as all recent handsets (Samsung S6 since 2015) include it as the opening of more 4G roaming agreements allow to use VoLTE in roaming.

If an MNO also operates fixed Internet services, IMS is the simplest and cheapest standard solution to allow using the same number in a mobile of fixed home environment. These services at home or mobile correspond simply to a change of S-CSCF and I-CSCF in the HLR-HSS which is common. This provides a standard replacement for the proprietary systems of the t which forwarded the calls to different forward or fixed numbers depending on the time or the result of the call. Table 7.1 shows the handling of the HLR-HSS for calls to a VoLTE handset and to a VoIP OTT handset mobile or fixed.

IMS is also the foundation of "Universal profile" services, previously called Rich Communication Suite (RCS) (see [0.9, Chapter 5]).

7.7 Roaming VoLTE with local break out

To improve the quality of outgoing calls in roaming, the UE will be authorized to use the PGW of the VPLMN in LBO mode and also the VPLMN P-CSCF. The rules so that the VPLMN accepts the LBO (the special APN such as "internet.eu" is not related to VoLTE; see [0.10, Chapter 8] for the control diagram).

7.8 Traces of user data (subscriber profile) in a server assignment answer (SAA/Cx)

In 4G, there is an UPDATE LOCATION sent by the MME. The answer ULA from the HLR-HSS contains the GPRS profile. In IMS, the equivalent is the server assignment request (SAR) sent by the lS-PCSF; the response SAA contains the VoIP profile of the subscriber with the MSISDN (public identity for VoTT VoIP or derived from the IMSI HLR-HSS interrogation) which will be used for the "SIP From" parameter of the outgoing SIP calls.

```
No.        Time        Source        Destination        Protocol Length Info
   49 14:32:51.052172000 172.27.40.1              172.27.10.1
DIAMETER/XML 2650    cmd=Server-AssignmentAnswer (301) flags=-P--
appl=3GPP Cx(16777216) h2h=3fc3650b e2e=23809a02


Frame 49: 2650 bytes on wire (21200 bits), 2650 bytes captured
(21200 bits) on interface 1
Ethernet II, Src: 00:00:00_00:00:00 (00:00:00:00:00:00),
Dst: 00:00:00_00:00:00 (00:00:00:00:00:00)
Internet Protocol Version 4, Src: 172.27.40.1 (172.27.40.1),
Dst: 172.27.10.1 (172.27.10.1)
Transmission Control Protocol, Src Port: diameter (3868),
Dst Port: 45156 (45156), Seq: 1, Ack: 525, Len: 2584
Diameter Protocol
    Version: 0x01
    Length: 2584
    Flags: 0x40
    Command Code: 301 Server-Assignment
    ApplicationId: 3GPP Cx (16777216)
    Hop-by-Hop Identifier: 0x3fc3650b
    End-to-End Identifier: 0x23809a02
    [Request In: 48]
    [Response Time: 0.011150000 seconds]
    AVP: Session-Id (263) l=65 f=-M- val=scscf-cxrf.epc.mnc001.mcc262.
3gppnetwork.org;1692213816;1
    AVP: Vendor-Specific-Application-Id (260) l=32 f=-M-
    AVP: Result-Code (268) l=12 f=-M- val=DIAMETER_SUCCESS (2001)
    AVP: Auth-Session-State (277) l=12 f=-M- val=NO_STATE_MAINTAINED
(1)
    AVP: Origin-Host (264) l=47 f=-M- val=hss-1.epc.mnc001.mcc262.
3gppnetwork.org
    AVP: Origin-Realm (296) l=41 f=-M- val=epc.mnc001.mcc262.
3gppnetwork.org
    AVP: User-Name (1) l=57 f=-M- val=262011234567890@ims.mnc001.
mcc262.3gppnetwork.org
    AVP: User-Data (606) l=2175 f=VM- vnd=TGPP
val=3c3f786d6c2076657273696f6e203d2022312e302220656e...
        AVP Code: 606 User-Data
        AVP Flags: 0xc0
        AVP Length: 2175
        AVP Vendor Id: 3GPP (10415)
        User-Data: 3c3f786d6c2076657273696f6e203d2022312e302220656e...
        eXtensible Markup Language
            <?xml
                version = "1.0"
                encoding = "UTF-8"
                ?>
            <IMSSubscription>
                <PrivateID>
                    262011234567890@ims.mnc001.mcc262.3gppnetwork.org
                    </PrivateID>
                <ServiceProfile>
```

```
            <PublicIdentity>  // IMSI
                <Identity>
                    sip:260011234567890@ims.
                    mnc001.mcc262.3gppnetworkorg // IMSI
                    </Identity>
                </PublicIdentity>
            <PublicIdentity>   // MSISDN
                <Identity>
                    tel:+491234567890     // MSISDN
                    </Identity>
                </PublicIdentity>
            <CoreNetworkServicesAuthorization>
                <SubscribedMediaProfileId>
                    0
                    </SubscribedMediaProfileId>
                </CoreNetworkServicesAuthorization>
          <Extension>
                <SharedIFCSetID>
                    1
                    </SharedIFCSetID>
                </Extension>
<InitialFilterCriteria>
    <Priority>0</Priority>
    <TriggerPoint>
      <ConditionTypeCNF>1</ConditionTypeCNF>
      <SPT>
      <ConditionNegated>0</ConditionNegated>
      <Group>0</Group>
      <Method>INVITE</Method>
      </SPT>
      <SPT>
       <ConditionNegated>0</ConditionNegated>
         <Group>0</Group>
      <Method>MESSAGE</Method>
      </SPT>
      <SPT>
       <ConditionNegated>0</ConditionNegated>
       <Group>0</Group>
       <Method>SUBSCRIBE</Method>
       </SPT>
       <SPT>
        <ConditionNegated>0</ConditionNegated>
        <Group>1</Group>
        <Method>INVITE</Method>
       </SPT>
       <SPT>
        <ConditionNegated>0</ConditionNegated>
        <Group>1</Group>
        <Method>MESSAGE</Method>
       </SPT>
       <SPT>
        <ConditionNegated>1</ConditionNegated>
        <Group>1</Group>
```

```
              <SIPHeader>
               <Header>From</Header>
               <Content>"TEST"</Content>
              </SIPHeader>
              </SPT>
            </TriggerPoint>
            <ApplicationServer>
                <ServerName>sip:AS1@italtel.com</ServerName>
            // for charging
                 <DefaultHandling>0</DefaultHandling>
             </ApplicationServer>
          </InitialFilterCriteria>
        </ServiceProfile>
      </IMSSubscription>
 AVP: Charging-Information (618) l=40 f=VM- vnd=TGPP
  AVP: Associated-Identities (632) l=72 f=V-- vnd=TGPP.
```

7.9 IMS files and certificates in the SIM card

7.9.1 IMS files in an ISIM card

To operate IMS, an MNO needs iSIM type 4G SIM cards; they include the directory ADMisim containing the IMS service parameters.

For 4G, the important files to include/modify are:

1. Files containing the PLMN with the technology information:

    - EFPLMNwACT (6F30)
    - EFOPLMNwACT (6F61)
    - EFHPLMNwACT (6F62)

2. Files for authentication and network selection

    - EFEPSLOCI (6FE3)
    - EFEPSNSC (6FE4)

3. Files for Emergency situations

    - EFICE_DN (6FE0)
    - EFICE_FF (6FE1)
    - EFICE_graphics (4F21)

4. Files for the network name

    - EFSPNI (6FDE)
    - EFPNNI (6FDF)

Below are the files for the IMS application [2.7]; some are permanent (Figure 7.11):

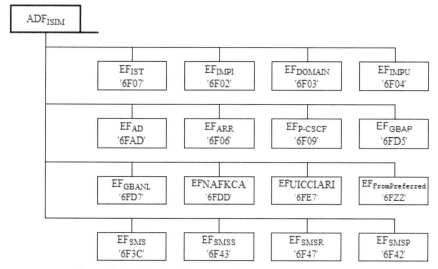

Figure 7.11 Standard SIM card files for the VoLTE services.

EFimpu contains the Public identity such as:

> sip:262011234567890@mnc001.mcc262.3gpp.org

EFp-CSCF contains the P-CSCF address to which is sent the SIP REGISTER

> sip:scscf.ims.mnc001.mcc262.3gppnetwork.org

EFist contains the IMS service list:

> Support of P-CSCF discovery for IMS Local Break Out, etc.

EFgbabp (GBA Bootstrapping parameters):

> AKA random challenge and Bootstrapping Transaction Identifier

Efsmsp (Short Message Service Parameters)

7.9.2 SIM files with the root certificate

For secured applications such as PMR, if a client VPN is used in the terminals, the access to services must use HTTPs, the same as Internet payments or bank account accesses. *In the servers,* there must be the X509 *client certificate* which must be allocated by a certification authority (CA) "root." For the terminal side, this certification must have been created by an AC the "root certificate" of which must be present in the terminal. The access is then secured by HTTPs/TLS with an RSA key of 1024 or 2048 bits (integer logarithm or better elliptic curves).

Manually by "rooting" the terminal, one can add the "root certificate," but this is impractical to do it on a number of terminals.It is quite more convenient to use the SIM card *to provide its own* "root certificate" for securing the access to the PMR servers using TLS.According to the 3GPP TS 31.102 standard:

1. The certificates are in the Directory DFMaxE
2. EForpk (Operator Root Public Key),
3. EFarpk (Administrator Root Public Key),
4. EFtprpk (Third Party Root Public Key).

Service 41 (MEXE) must be activated in the SST profile. This can be done by OTA. The terminals must be capable of using these additional root certificates in the SIM card which are read in the SIM card when the UE is powered up or receives a REFRESH command sent by a SIM STK when it receives an OTA refresh command.

If the MNO operator is happy with certificates signed by a chain the root of which is a third party, foreign in particular, he does not need to provide another root certificate to certify its own. There is no need for the above then.

The use of dubious root certificates is something to avoid; they must be deleted in any secure handset; Google can geoloc most of the terminals in the world because there supl.google.com certificate is included in practically all the Androd and Iphone handsets, which is a risk (see [0.10, Chapter 12] for an attack example).

7.10 QoS parameter mapping GTPv2 ← Gx ← Rx ← application function

This is described in TS 29.213 "LTE; policy and charging control signaling flows and quality of service (QoS) parameter mapping" when the P-CSCF sends an AA Request/Gx to the PCRF which will ask to create a new dedicated bearer. It describes what the PCRF does on this example of a bearer for an AUDIO codec.

7.10.1 Dedicated bearer creation, traces Rx and Gx

Here are the traces of a dialog between the core IMS and the PGW. The P-CSCF provides the PCRF with parameters describing an AUDIO media. The

PCRF uses the 3GPP rule and the subscriber profile identified in the Credit Control Request (CCR) from the PCEF. It uses this identifier to obtain the subscriber profile in its database as received from the UDR in Figure 7.1. A Re-Auth Request/Gx is sent to the PCEF which will create a "dedicated bearer" with the "Traffic Templates" from the P-CSCF and the QoS computed by the PCRF.

```
No.        Time               Source          Destination    Protocol Length
  2635 11:02:36.892965000 192.168.2.114 192.168.2.116 DIAMETER 778
Info cmd=AA Request (265) flags=R--- appl=3GPP
Rx(16777236) h2h=1c1924 e2e=3c51d2d7 |


Frame 2635: 778 bytes on wire (6224 bits), 778 bytes captured
(6224 bits) on interface 0
Ethernet II, Src: Vmware_af:1c:2a (00:0c:29:af:1c:2a),
Dst: Vmware_ea:40:0f (00:0c:29:ea:40:0f)
Internet Protocol Version 4, Src: 192.168.2.114 (192.168.2.114),
Dst: 192.168.2.116 (192.168.2.116)
Stream Control Transmission Protocol, Src Port: 39144 (39144),
Dst Port: 3868 (3868)
Diameter Protocol
    Version: 0x01
    Length: 716
    Flags: 0x80
    Command Code: 265 AA
    ApplicationId: 3GPP Rx (16777236)
    Hop-by-Hop Identifier: 0x001c1924
    End-to-End Identifier: 0x3c51d2d7
    [Answer In: 2637]
    AVP: Session-Id (263) l=44 f=-M- val=pcscf8.corp.cirpack.com;
    1482294574;2
    AVP: Auth-Application-Id (258) l=12 f=-M- val=3GPP Rx (16777236)
    AVP: Origin-Host (264) l=23 f=-M- val=af_dev.halys.fr
    AVP: Origin-Realm (296) l=16 f=-M- val=halys.fr
    AVP: Destination-Realm (283) l=16 f=-M- val=halys.fr
    AVP: Destination-Host (293) l=21 f=-M- val=pcrf.halys.fr
    AVP: AF-Application-Identifier (504) l=18 f=VM- vnd=TGPP
    val=706373636638
    AVP: Media-Component-Description (517) l=472 f=VM- vnd=TGPP
        AVP Code: 517 Media-Component-Description
        AVP Flags: 0xc0
        AVP Length: 472
        AVP Vendor Id: 3GPP (10415)
        Media-Component-Description: 00000206c0000010000028af0000000
        100000207c00000ac...
            AVP: Media-Component-Number (518) l=16 f=VM- vnd=TGPP
            val=1
            AVP: Media-Sub-Component (519) l=172 f=VM- vnd=TGPP
                AVP Code: 519 Media-Sub-Component
                AVP Flags: 0xc0
```

```
                    AVP Length: 172
                    AVP Vendor Id: 3GPP (10415)
                    Media-Sub-Component: 000001fdc0000010000028af0000000
                    1000001fbc0000047...
                        AVP: Flow-Number (509) l=16 f=VM- vnd=TGPP
                        val=1
                        AVP: Flow-Description (507) l=71 f=VM- vnd=TGPP
val=permit out 17 from 192.168.37.87 30156 to 199.255.7.39 6350
                        AVP: Flow-Description (507) l=70 f=VM- vnd=TGPP
val=permit in 17 from 199.255.7.39 6350 to 192.168.37.87 30156
                AVP: Media-Sub-Component (519) l=188 f=VM- vnd=TGPP
                    AVP Code: 519 Media-Sub-Component
                    AVP Flags: 0xc0
                    AVP Length: 188
                    AVP Vendor Id: 3GPP (10415)
                    Media-Sub-Component: 000001fdc0000010000028af0000000
                    2000001fbc0000047...
                        AVP: Flow-Number (509) l=16 f=VM- vnd=TGPP val=2
                        AVP: Flow-Description (507) l=71 f=VM- vnd=TGPP
val=permit out 17 from 192.168.37.87 30157 to 199.255.7.39 6351
                        AVP: Flow-Description (507) l=70 f=VM- vnd=TGPP
val=permit in 17 from 199.255.7.39 6351 to 192.168.37.87 30157
                        AVP: Flow-Usage (512) l=16 f=VM- vnd=TGPP
val=RTCP (1)
                    AVP: AF-Application-Identifier (504) l=18 f=VM- vnd=TGPP
        val=706373636638
                    AVP: Media-Type (520) l=16 f=VM- vnd=TGPP
val=AUDIO (0)
                    AVP: Max-Requested-Bandwidth-UL (516) l=16
f=VM- vnd=TGPP val=64000
                    AVP: Max-Requested-Bandwidth-DL (515) l=16
f=VM- vnd=TGPP val=64000
                    AVP: Flow-Status (511) l=16 f=VM- vnd=TGPP val=ENABLED (2)
        AVP: Supported-Features (628) l=56 f=VM- vnd=TGPP
            AVP Code: 628 Supported-Features
            AVP Flags: 0xc0
            AVP Length: 56
            AVP Vendor Id: 3GPP (10415)
            Supported-Features: 0000010a4000000c000028af000002758000001
            0000028af...
                AVP: Vendor-Id (266) l=12 f=-M- val=10415
                AVP: Feature-List-ID (629) l=16 f=V-- vnd=TGPP val=1
                AVP: Feature-List (630) l=16 f=V-- vnd=TGPP val=19
        AVP: Framed-IP-Address (8) l=12 f=-M- val=199.255.7.39
        (199.255.7.39)

No.      Time              Source       Destination    Protocol Length
   2637 11:02:36.916598000 192.168.2.116 192.168.2.114 DIAMETER 238
Call-ID Info SACK cmd=AA Answer (265) flags=----- appl=3GPP
Rx(16777236) h2h=1c1924 e2e=3c51d2d7 |

Frame 2637: 238 bytes on wire (1904 bits), 238 bytes captured
(1904 bits) on interface 0
```

```
Ethernet II, Src: Vmware_ea:40:0f (00:0c:29:ea:40:0f),
Dst: Vmware_af:1c:2a (00:0c:29:af:1c:2a)
Internet Protocol Version 4, Src: 192.168.2.116 (192.168.2.116),
Dst: 192.168.2.114 (192.168.2.114)
Stream Control Transmission Protocol, Src Port: 3868 (3868),
Dst Port: 39144 (39144)
Diameter Protocol
    Version: 0x01
    Length: 160
    Flags: 0x00
    Command Code: 265 AA
    ApplicationId: 3GPP Rx (16777236)
    Hop-by-Hop Identifier: 0x001c1924
    End-to-End Identifier: 0x3c51d2d7
    [Request In: 2635]
    [Response Time: 0.023633000 seconds]
    AVP: Session-Id (263) l=44 f=-M- val=pcscf8.corp.cirpack.com;
    1482294574;2
    AVP: Auth-Application-Id (258) l=12 f=-M- val=3GPP Rx (16777236)
    AVP: Origin-Host (264) l=21 f=-M- val=pcrf.halys.fr
    AVP: Origin-Realm (296) l=16 f=-M- val=halys.fr
    AVP: Result-Code (268) l=12 f=-M- val=DIAMETER_SUCCESS (2001)
    AVP: IP-CAN-Type(1027) l=16 f=VM- vnd=TGPP val=3GPP-GPRS (0)
    AVP: RAT-Type(1032) l=16 f=V-- vnd=TGPP val=HSPA_EVOLUTION (1003)
// le PCRF le sait du CCR de la création de session.

No.      Time        Source         Destination      Protocol Length Info
     3 12:02:36.926372   192.168.2.116        192.168.2.114
DIAMETER 942    cmd=Re-Auth Request (258) flags=R--- appl=3GPP
Gx(16777238) h2h=1c1924 e2e=3c51d2d7 |

Frame 3: 942 bytes on wire (7536 bits), 942 bytes captured
(7536 bits) on interface 0
Ethernet II, Src: Vmware_ea:40:0f (00:0c:29:ea:40:0f),
Dst: Vmware_af:1c:2a (00:0c:29:af:1c:2a)
Internet Protocol Version 4, Src: 192.168.2.116,
Dst: 192.168.2.114
Stream Control Transmission Protocol, Src Port: 3868 (3868),
Dst Port: 39144 (39144)
Diameter Protocol
    Version: 0x01
    Length: 880
    Flags: 0x80, Request
    Command Code: 258 Re-Auth
    ApplicationId: 3GPP Gx (16777238)
    Hop-by-Hop Identifier: 0x001c1924
    End-to-End Identifier: 0x3c51d2d7
    AVP: Session-Id (263) l=12 f=-M- val=7890
    AVP: Auth-Application-Id (258) l=12 f=-M- val=3GPP Gx (16777238)
    AVP: Origin-Host (264) l=21 f=-M- val=pcrf.halys.fr
    AVP: Origin-Realm (296) l=16 f=-M- val=halys.fr
    AVP: Destination-Realm (283) l=16 f=-M- val=halys.fr
    AVP: Destination-Host (293) l=25 f=-M- val=pgw_pcef.halys.fr
```

```
AVP: Re-Auth-Request-Type (285) l=12 f=-M- val=AUTHORIZE_ONLY (0)
AVP: Charging-Rule-Install(1001) l=740 f=VM- vnd=TGPP
    AVP Code: 1001 Charging-Rule-Install
    AVP Flags: 0xc0
    AVP Length: 740
    AVP Vendor Id: 3GPP (10415)
    Charging-Rule-Install: 000003ebc000016c000028af000003edc00000
    10000028af...
        AVP: Charging-Rule-Definition(1003) l=364 f=VM- vnd=TGPP
            AVP Code: 1003 Charging-Rule-Definition
            AVP Flags: 0xc0
            AVP Length: 364
            AVP Vendor Id: 3GPP (10415)
            Charging-Rule-Definition: 000003edc0000010000028
            af0001000100000042280000054...
                AVP: Charging-Rule-Name(1005) l=16 f=VM-
                vnd=TGPP val=00010001
                AVP: Flow-Information(1058) l=84 f=V-- vnd=TGPP
                    AVP Code: 1058 Flow-Information
                    AVP Flags: 0x80
                    AVP Length: 84
                    AVP Vendor Id: 3GPP (10415)
                    Flow-Information: 000001fbc0000047000028
                    af7065726d6974206f75742031...
                        AVP: Flow-Description (507) l=71 f=VM-
vnd=TGPP val=permit out 17 from 192.168.37.87 30156 to 199.255.7.39
6350
                AVP: Flow-Information(1058) l=84 f=V-- vnd=TGPP
                    AVP Code: 1058 Flow-Information
                    AVP Flags: 0x80
                    AVP Length: 84
                    AVP Vendor Id: 3GPP (10415)
                    Flow-Information: 000001fbc0000046000028
                    af7065726d697420696e203137...
                        AVP: Flow-Description (507) l=70 f=VM-
vnd=TGPP val=permit in 17 from 199.255.7.39 6350 to 192.168.37.87
30156
                AVP: Flow-Status (511) l=16 f=VM- vnd=TGPP
                val=ENABLED (2)
                AVP: QoS-Information(1016) l=152 f=VM- vnd=TGPP
                    AVP Code: 1016 QoS-Information
                    AVP Flags: 0xc0
                    AVP Length: 152
                    AVP Vendor Id: 3GPP (10415)
                    QoS-Information: 00000404c0000010000028
                    af0000000100000204c0000010...
                        AVP: QoS-Class-Identifier(1028) l=16
                        f=VM- vnd=TGPP val=QCI_1 (1)
                        AVP: Max-Requested-Bandwidth-UL (516) l=16
                        f=VM- vnd=TGPP val=64000
                        AVP: Max-Requested-Bandwidth-DL (515) l=16
                        f=VM- vnd=TGPP val=64000
                        AVP: Guaranteed-Bitrate-UL(1026) l=16
```

```
                              f=VM- vnd=TGPP val=64000
                              AVP: Guaranteed-Bitrate-DL(1025) l=16
                              f=VM- vnd=TGPP val=64000
Set by the PCRF depending on the subscriber profile: RTP
(voice packets):
          AVP: Allocation-Retention-Priority(1034)
              l=60 f=V-- vnd=TGPP
                  AVP Code: 1034 Allocation-Retention
                   -Priority
                  AVP Flags: 0x80
                  AVP Length: 60
                  AVP Vendor Id: 3GPP (10415)
                  Allocation-Retention-Priority:
                  00000416800000010000028af000000010000041780000010...
                      AVP: Priority-Level(1046) l=16
                       f=V-- vnd=TGPP val=1
                      AVP: Pre-emption-Capability(1047)

    l=16 f=V-- vnd=TGPP val=PRE-EMPTION_CAPABILITY_ENABLED (0)
                      AVP: Pre-emption-Vulnerability

   (1048) l=16 f=V-- vnd=TGPP val=PRE-EMPTION_VULNERABILITY_DISABLED (1)

          AVP: Charging-Rule-Definition(1003) l=364 f=VM- vnd=TGPP
              AVP Code: 1003 Charging-Rule-Definition
              AVP Flags: 0xc0
              AVP Length: 364
              AVP Vendor Id: 3GPP (10415)
              Charging-Rule-Definition: 000003edc0000010000028
af00010002000000042280000054...
                  AVP: Charging-Rule-Name(1005) l=16 f=VM-
vnd=TGPP val=00010002
                  AVP: Flow-Information(1058) l=84 f=V-- vnd=TGPP
                      AVP Code: 1058 Flow-Information
                      AVP Flags: 0x80
                      AVP Length: 84
                      AVP Vendor Id: 3GPP (10415)
                      Flow-Information: 000001fbc0000047000028
af7065726d6974206f75742031...
                          AVP: Flow-Description (507) l=71 f=VM-
vnd=TGPP val=permit out 17 from 192.168.37.87 30157 to 199.255.7.39
6351
                  AVP: Flow-Information(1058) l=84 f=V-- vnd=TGPP
                      AVP Code: 1058 Flow-Information
                      AVP Flags: 0x80
                      AVP Length: 84
                      AVP Vendor Id: 3GPP (10415)
                      Flow-Information: 000001fbc0000046000028
af7065726d697420696e203137...
                          AVP: Flow-Description (507) l=70 f=VM-
vnd=TGPP val=permit in 17 from 199.255.7.39 6351 to 192.168.37.87
30157
                  AVP: Flow-Status (511) l=16 f=VM- vnd=TGPP
```

```
                        val=ENABLED (2)
                AVP: QoS-Information(1016) l=152 f=VM- vnd=TGPP
                    AVP Code: 1016 QoS-Information
                    AVP Flags: 0xc0
                    AVP Length: 152
                    AVP Vendor Id: 3GPP (10415)
                    QoS-Information: 00000404c0000010000028
                    af0000000100000204c0000010...
                        AVP: QoS-Class-Identifier(1028) l=16 f=VM-
                        vnd=TGPP val=QCI_1 (1)
                        AVP: Max-Requested-Bandwidth-UL (516) l=16
                        f=VM- vnd=TGPP val=3200
                        AVP: Max-Requested-Bandwidth-DL (515) l=16
                        f=VM- vnd=TGPP val=3200
                        AVP: Guaranteed-Bitrate-UL(1026) l=16
                        f=VM- vnd=TGPP val=3200
                        AVP: Guaranteed-Bitrate-DL(1025) l=16
                        f=VM- vnd=TGPP val=3200
Set by the PCRF depending on the subscriber profile: RTCP
            AVP: Allocation-Retention-Priority(1034)
                l=60 f=V-- vnd=TGPP
                    AVP Code: 1034 Allocation-Retention-
                    Priority
                    AVP Flags: 0x80
                    AVP Length: 60
                    AVP Vendor Id: 3GPP (10415)
                    Allocation-Retention-Priority:
                    0000041680000001000028af0000000010000041780000010...
                        AVP: Priority-Level(1046)
                        l=16 f=V-- vnd=TGPP val=1
                        AVP: Pre-emption-Capability
(1047) l=16 f=V-- vnd=TGPP val=PRE-EMPTION_CAPABILITY_ENABLED (0)
                        AVP: Pre-emption-Vulnerability
(1048) l=16 f=V-- vnd=TGPP val=PRE-EMPTION_VULNERABILITY_DISABLED (1)
```

7.10.2 PCRF processing: Correspondences between the GTPv2 parameters and the AVPs Gx et Rx

To simplify, we assume that only one "dedicated bearer" is created for RTP. Above the P-CSCF had described two media: RTP and RTCP. The PGW receives "flow descriptions" from the PCRF and computes traffic flow templates (TFTs). These are rules (IPs and ports) allowing the PCEF to route the DL trafic to the dedicated bearer which is created and the UE to route the UL traffic to the particular bearer. They are shown with a gray back plane.

GTP v2 parameter in Create Bearer Request (set by the PCEF)	Gx AVP in Re-Auth Request (set by PCRF)	Rx AVP in AA Request set by the P-CSCF
On ne crée un Bearer que pour RTP ci-cessous **84):EPS Bearer Level Traffic Flow Template(TFT)** TFT operation: (1):Create new TFT E bit: (0):E-bit: parameter list not included Number of packet filters: 2 -Packet Filter direction 1: (2):uplink only -Packet Filter identifier 1: 1 -Packet Filter evaluation precedence 1: 41(29) -Packet filter 1 length: 28 -Packet filter component type identifier: (16):IPv4 remote address type IPv4 addr: **199.255.7.39** addresse de l'UE 'distante' IPv4 mask: 255.255.255.255 -Packet filter component type identifier: (17):IPv4 **local** address type IPv4 addr: 199.255.7.39 addresse de l'**UE auquel on ajoute ce Bearer dédié** IPv4 mask: 255.255.255.255 -Packet filter component type identifier: (64):Single local port type Port: **30156** -Packet filter component type identifier: (80):Single Remote port type Port: **6350** -Packet filter component type identifier: (128):Flow label type Flow label type: 000001	Charging-Rule-Definition: 000003edc0000010000028af 00010001000004228000 0054... AVP: Charging-Rule-Name(1005) l=16 f=VM- vnd=TGPP val=00010001 AVP: Flow-Information(1058) l=84 f=V-- vnd=TGPP AVP Code: 1058 Flow-Information AVP Flags: 0x80 AVP Length: 84 AVP Vendor Id: 3GPP (10415) Flow-Information: 000001fbc0000047000028 af7065726d6974206f7574 2031... AVP: Flow-Description (507) l=71 f=VM- vnd=TGPP val=permit out 17 from 192.168.37.87 30156 to **199.255.7.39 6350** AVP: Flow-Information(1058) l=84 f=V-- vnd=TGPP AVP Code: 1058 Flow-Information AVP Flags: 0x80 AVP: Flow-Information(1058) l=84 f=V-- vnd=TGPP AVP Code: 1058 Flow-Information AVP Flags: 0x80 AVP Length: 84 AVP Vendor Id: 3GPP (10415) Flow-Information: 000001fbc0000046000028 af7065726d697420696e20 3137... AVP: Flow-Description (507) l=70 f=VM- vnd=TGPP val=permit in 17 from 199.255.7.39 6350 to 192.168.37.87 **30156**	Media-Sub-Component: 000001fdc0000010000028 af00000001000001fbc0000 047... AVP: Flow-Number (509) l=16 f=VM- vnd=TGPP val=1 AVP: Flow-Description (507) l=71 f=VM- vnd=TGPP val=permit out 17 from 192.168.37.87 30156 to 199.255.7.14 6350 AVP: Flow-Description (507) l=70 f=VM- vnd=TGPP val=permit in 17 from 199.255.7.39 6350 to 192.168.37.87 30156 (No AVP: Flow-Usage → default (RTP)) Media-Sub-Component: 000001fdc0000010000028af 00000002000001fbc0000 047... AVP: Flow-Number (509) l=16 f=VM- vnd=TGPP val=2 AVP: Flow-Description (507) l=71 f=VM- vnd=TGPP val=permit out 17 from 192.168.37.87 30157 to 199.255.7.14 6351 AVP: Flow-Description (507) l=70 f=VM- vnd=TGPP val=permit in 17 from 199.255.7.39 6351 to 192.168.37.87 30157 AVP: Flow-Usage (512) l=16 f=VM- vnd=TGPP val=RTCP (1)

GTP v2 parameter in Create Bearer Request (set by the PCEF)	Gx AVP in Re-Auth Request (set by PCRF)	Rx AVP in AA Request set by the P-CSCF
		AVP: Framed-IP-Address (8) l=12 f=-M- val=199.255.7.39 (199.255.7.39)
(80):Bearer level Quality of Service (flow RTP) ARP(Allocation and Retention Priority): Priority Level: 1 Pre-emption Capability: (0):Enabled Pre-emption Vulnerability: (1):Disabled QCI(QoS Class of Identifier): (1):GBR,priority = 2, 100 ms: VoIP Maximum Bit Rate for Uplink: 64 000 bps Maximum Bit Rate for Downlink: 64000 bps Guaranteed Bit Rate for Uplink(5 oct): 30000 kbps Guaranteed Bit Rate for Downlink(5 oct): 30000 kbps	Setting of the QCI and QoS 1) PCRF uses first TS 29.213 to derive the QCI and the bandwidths standard for the codecs. 2) Then the PCRF uses the subscriber profile to have the customized QoS for the same QCI and set the ARP (priority and pre-emptivity) AVP: QoS-Information(1016) l=92 f=VM- vnd=TGPP AVP: QoS-Class-Identifier(1028) l=16 f=VM- vnd=TGPP val=QCI_1 (1) AVP: Max-Requested-Bandwidth-UL (516) l=16 f=VM- vnd=TGPP val=64000 AVP: Max-Requested-Bandwidth-DL (515) l=16 f=VM- vnd=TGPP val=64000 AVP: Guaranteed-Bitrate-UL(1026) l=16 f=VM- vnd=TGPP val=64000 AVP: Guaranteed-Bitrate-DL(1025) l=16 f=VM- vnd=TGPP val=64000 AVP: Allocation-Retention-Priority(1034) l=60 f=V-- vnd=TGPP AVP Code: 1034 Allocation-Retention-Priority AVP Flags: 0x80 AVP Length: 60 AVP Vendor Id: 3GPP (10415)	AVP: Media-Type (520) l=16 f=VM- vnd=TGPP val=AUDIO (0) AVP: Max-Requested-Bandwidth-UL (516) l=16 f=VM- vnd=TGPP val=64000 AVP: Max-Requested-Bandwidth-DL (515) l=16 f=VM- vnd=TGPP val=64000

GTP v2 parameter in Create Bearer Request (set by the PCEF)	Gx AVP in Re-Auth Request (set by PCRF)	Rx AVP in AA Request set by the P-CSCF
	Allocation-Retention-Priority: 0000043600000100008295f0 0000001000094178000010,... AVP: Priority-Level (1046) 1=16 S-V vnd-3GPP val-1 AVP: Pre-emption-Capability (1047) 1=16 f-V vnd-3GPP val=PRE-EMPTION_CAPABILITY_ ENABLED (0) AVP: Pre-emption-Vulnerability (1048) 1=16 f-V- vnd-3GPP val=PRE_EMPTION_VULNERABILITY_ DISABLED (1)	

7.11 Access through a non-trusted WLAN (WiFi) to the PMR core networking

7.11.1 Operational interest

Public security forces, police, and firemen are often stationed in fixed offices from which they intervene. It is important that their internal communications including office–office be secured and cheap and hence the use of WiFi base stations which should be able to be used as a non-trusted access to the PMR core network AS for SIP-based call making and receiving. The 3GPP standard architecture allows an economic implementation of this service using basic WiFi stations which *do not need to implement EAP-SIM Radius as it uses an IPsec tunnel* between the UE and the ePDG. The only assumption in the scheme described, there are others, is that the UEs must have a WiFi client which can access the SIM card to implement the EAP authentication scheme. Among the MNOs, VoWiFi has become popular as the VoWiFI client is available in the recent Androïd or IoS handsets with certificates locking to the HPLMN. The service can be tested in a lab with a standard client emulator which does use certificates.

7.11.2 Call flow to establish the IPsec tunnel and localize a WiFi UE in the HLR-HSS

The full IKE/EAP-AKA/ESP flow is shown in Figure 7.12. Internet Key Exchange (IKE) is the protocol used between the handset (WiFi connected) and the ePDG server to set the shared security information to set an Ipsec

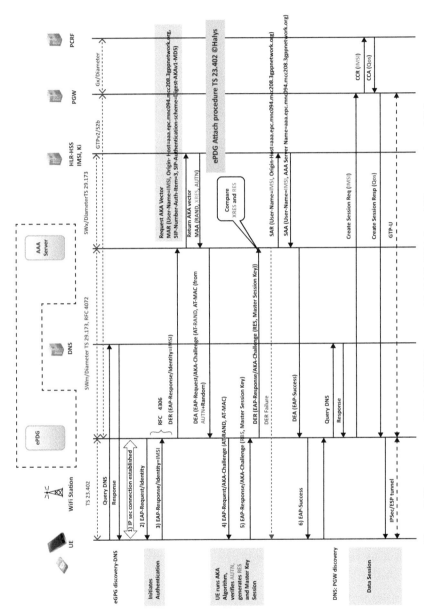

Figure 7.12   Non-trusted WLAN WiFI access with EAP-AKA based IPsec tunnel establishment.

Table 7.5   Detailed messages of the SWm/Diameter messages

Exchange RFC 4306 (Internet Key Exchange)	SWm/ Diameter TS 29.273 and RFC 4072	SWx /Diameter TS 29.273	Comparison with EAP-AKA 802.1x using RADIUS [0.9, page 141] RFC 5448 (implementation EAP-AKA) RFC 4186 (implementation
Handset → ePDG	ePDG → AAA server	AAA server → HSS	EAP-SIM WiFi station → AAA server
	DER		Access-Request/EAP/Response/ IMSI →
	DEA		Access-Challenge/EAP/Request/EAP-AKA/Start
	DER		Access-Request/EAP/Response/EAP-AKA/Start NONCE, Version →
	DEA		Access-Challenge/EAP/Request/EAP-AKA/Challenge(RAND, MAC)+Start →
	DER		Access-Request/EAP/Response/EAP-AKA/
			Comparison by the AAA of the computations by the SIM and by the HLR-HSS of the XRES and RES values

tunnel. IKE is now defined in RFC 4306 [7.25] as IKEv2. It uses the Diffie–Hellman (DH) exchange to set a shared secret, which is used to derive the session ciphering keys.

Figure 7.12 and the Table 7.5 gives the details of the messages on the SWm interface.

IKE is used to identify both ends (the handset with "supplicant" provides the IMSI which it must be able to access in the SIM card). Two authentication schemes are possible: PSK to generate RSA session keys or with certificates issued by a CA:

A reason to describe the non-trusted access architecture in the same chapter as IMS is the SWx/Diameter protocol as similar concepts with Cx/Diameter such as the SAR message which will allow the reception of calls to the WiFi UE because the HLR-HSS knows which AAA server the user is visiting. In the main mode, the initiator UE and recipient send three two-way exchanges (six messages total) to accomplish the following services:

- First exchange (messages 1 and 2) – Proposes and accepts the encryption and authentication algorithms.
- Second exchange (messages 3 and 4) – Executes a DH exchange, and the initiator and recipient each provide a pseudorandom number.
- Third exchange (messages 5 and 6) – Sends and verifies the identities of the initiator (the IMSI) and the recipient.

The information transmitted in the third exchange of messages (after RUN AKA in Figure 7.12) is protected by the encryption algorithm established in the first two exchanges. Thus, the participants' identities are encrypted and therefore not transmitted "in the clear."

- Diameter-EAP-Request (DER) includes the IMSI in the AVP user name.
- Diameter-EAP-Answer includes the AVP EAP-payload used to challenge the SIM card.

The IKE or IPSec SAs use secret keys that should be used for a limited time and to protect a limited amount of data. This is because we want to make sure that even if an attacker finds out the secret keys (e.g., by using some brute force mechanism), the amount of data compromised is limited. After the IKE or IPSec SAs is expired, a new security association is established to take place of the expired one. The process is known as "Rekeying."

The rekeying process could be triggered based on the duration or amount of data transferred on the existing IPSec tunnel and can be initiated by both ePDG and IKEv2 peer. The rekeying is done using a CREATE_CHILD_SA exchange. If both the IKE and IPSec security association require rekeying, then they are performed separately.

For EAP-SIM/AKA, we have an optional but important optimization called *fast re-authentication*. It is defined by the GSMA IR.61 and TS 33.402 [7.27]. It provides authentication that does not require new vectors from the HLR/HSS. The original master session key (MSK) from the full authentication is used to generate a fresh MSK. That means that the new triplets from the HSS are not necessary. The UE uses the fast re-authentication identity returned by the AAA. Naturally, the fast re-authentication reduces the load on the HLR/HSS. The "Authentication-Scheme" for SWx[7.23] is the same as for Cx [7.11] used in IMS including the basic authentication with ID and password but also EAP-AKA [7.27].

References

[7.1] 3GPP TS 24.611 V12.4.0 (2015-01), "Digital cellular telecommunications system (Phase 2+); Universal Mobile Telecommunications System (UMTS); LTE; Anonymous Communication Rejection (ACR) and Communication Barring (CB) using IP Multimedia (IM) Core Network (CN) subsystem; Protocol specification (3GPP TS 24.611 version 12.4.0 Release 12)," *description of the HTTP XML protocol for the supplementary services between the UE and the AS.*

[7.2] 3GPP TS 24.623 v13.1.0 (2016-01), "Digital cellular telecommunications system (Phase 2+); Universal Mobi le Telecommunications System (UMTS); LTE; Extensible Markup Language (XML) Configuration Access Protocol (XCAP) over the Ut interface for Manipulating Supplementary Services."

[7.3] GSMa NG103, "VoLTE RCS Roaming and Interconnection Guidelines," Version 1.0, 19 May 2015.

[7.4] GSMa PRD IR.88, "LTE and EPC Roaming Guidelines," Version 11.0, 20 January 2014.

[7.5] GSMa PRD IR.65, "IMS Roaming & Interworking Guidelines," Version 14.0, 28 April 2014.

[7.6] www.kamailio.org, Open source SIP server, *successor of Open SIP, v4.4.5, January, has the Diameter interfaces for IMS including the prepaid billing Ro/Diameter.*

[7.7] 3GPP TS 31.103, v13.1.0, Release 13, "LTE; Characteristics of the IP Multimedia Services Identity Module (ISIM) application."

[7.8] IETF Draft, "Prepaid extensions to remote authentication Dial-in user service (RADIUS)," February 2013.

[7.9] 3GPP TS 29.274, v12.8.0, Release 12, "LTE, General Tunnelling Protocol for Control Plane (GTP v2 Stage 3," *protocol GTP v2 between MME and S/PGW.*

[7.10] RFC 7315, "Private Header (P-Header) Extensions to the Session Initiation Protocol (SIP) for the 3GPP," July 2014.

[7.11] 3GPP TS 29.229 v14.1.0 (2017-04), "Digital cellular telecommunications system (Phase 2+) (GSM); Universal Mobile Telecommunications System (UMTS); LTE; Cx and Dx interfaces based on the Diameter protocol; Protocol details," Release 14. *Interface Cx between I-CSCF or S-CSCF and HSS.*

[7.12] RFC 2069, "An extension to HTTP: Digest Access Authentication," Jan 1997. *The "response" =MD5 (MD5(user:realm:password):Nonce: MD5 (method:Digest-URI)).*

[7.13] "IMS Profile for Voice and SMS Version," GSMa IR 92, v7.0 03, March 2013.

[7.14] RFC 2617, "HTTP Authentication: Basic and Digest Access Authentication," June 1999, Details for the coding of Digest-QoP and Digest-HA1 in the Multimedia Authentication Answer using SIP Digest (VOTT VoIP).

[7.15] 3GPP TS 23.401 V14.3.0 (2017-05), "LTE; General Packet Radio Service (GPRS) enhancements for Evolved Universal Terrestrial Radio Access Network (E-UTRAN) access," *Annex K describes the IOPS architecture.*

[7.16] A. Henry-Labordère, "Système Optimisant le Renvoi Tardif d'Appels vers une messagerie vocale de mobile (SORTA)," Patent FR 05 51804.

[7.17] Q-763 Release 1999,ITU-T, "Signalling System N°7, ISDN User Part (ISUP), *formats and codes, All the detailed ISUP messages.*

[7.18] ARCEP, Document ARCEF/SFM/UAI/09-2005, *Particularities of the french "SPIROU" ISUP.*

[7.19] 3GPP TS 29.272 V14.5.0 (2017-10), "Universal Mobile Telecommunications System (UMTS); LTE; Evolved Packet System (EPS); Mobility Management Entity (MME) and Serving GPRS Support Node (SGSN) related interfaces based on Diameter protocol," Release 14, *The S6a/S6b and S13 interfaces to the HSS from a MME or SGSN.*

[7.20] 3GPP TS 29.212 V14.5.0 (2017-10), "Universal Mobile Telecommunications System (UMTS); LTE; Policy and Charging Control (PCC); Reference points," Release 14, *the Gx interface between the PGW and the PCRF.*

[7.21] TS 29.214 V14.5.9 (2017-10), "Universal Mobile Telecommunications System (UMTS); LTE; Policy and charging control over Rx reference point," Release 14, *the Rx interface between the PCRF and the Application servers (AS).*

[7.22] TS 23.402 V14.6.0 (2017), "Universal Mobile Telecommunications System (UMTS); LTE; Architecture enhancements for non-3GPP accesses," Release 14. *Describes the ePDG, the AAA server, and the protocols.*

[7.23] TS 29.273 V13.2.0 (2016-10) "Universal Mobile Telelecommunications System (LTE; Evolved Packet System (EPS); 3GPP EPS AAA interfaces," Release 13, *the SWu, SWm, SWx Diameter protocols.*

[7.24] StrongSWAN

[7.25] RFC 4306, "Internet Key Exchange (IKEv2) Protocol," December 2005

[7.26] RFC 4072, "Diameter Extensible Authentication Protocol (EAP) Application," 2005, *between ePDG and AAA server, same as SWm.*

[7.27] TS 33.402 V14.3.0 (2017-10), "Digital cellular telecommunications system (Phase 2+) (GSM); Universal Mobile Telecommunications System (UMTS); LTE; 3GPP System Architecture Evolution (SAE); Security aspects of non-3GPP accesses," Release 14.

8

Lawful Interception 3GPP Architecture and PMR Network Case

$$\frac{\pi^2}{6} = 1 + \frac{1}{n} + \frac{1}{n^2} + \frac{1}{n^3} + \frac{1}{n^4} \cdots$$

I discovered unexpectedly an elegant expression of the sum of the serie: I discovered that six times the sum is equal to the square of the length of the circle with diameter equal to 1.
Leonhard Euler, solution to the Bale problem (1741).

8.1 Legal interception applied to PMR networks: Use for monitoring and security

The PMR also needs the legal interception: as a log of the interventions and as a way of controlling the security force members, army, police, and firefighting forces may be the subject of legal interceptions at all time including when they are under the PMR network coverage. This is often forgotten in the design of PMR networks; LI support should be provided.

8.2 The LI standard 3GPP architecture

8.2.1 Proprietary interfaces of the network equipment and standardization

In countries where the MNOs connected to the public network are required to provide the legal interception services to the LEA, this is a costly addition to their core network, if the cost is not born by the LEA. The LEA requires the compliance with standard interfaces to their LEMF, while the equipment in the core network may come from different vendors. So, in practice, a mediation system will need to be used which has both.

- the newly standardized X1 (2017) [8.5] for the lawful interception commands;
- the still proprietary X2 and X3 interfaces with the MNO's equipment; [8.5] will have Parts 2 and 3 to standardize them;
- the standard interfaces HI1, HI2, and HI3 with the LEMF.

Figure 8.1 illustrates an architecture using a mediation system to interconnect to the LEA's LEMF.

In the core network's LI implementation, there is a "lawful intercept protected proxy" (colored in "Kawasaki green") which interfaces with the "Mediation Function" in orange using HTTPs in both ways. This way the data are protected end-to-end between X2 intercept and X3 intercept HTTPS client and the MF. Also the identities of the intercepted persons are hidden by HTTPs for X1 while transmitted, and ciphered in the database of the lawful intercept protected proxy. Following security regulations of some agencies such as ANSSI (France) R226, it protects personal information from fraudulent external crashes or dumps provoked by operation staffs. No one should ever be able to obtain the IDs of the intercepted persons even from inside the MNO.

In this implementation example, all the data, intercepted or not, are sent to the "proxy" which selects in real time those which are sent to the MF system. As seen in Figure 8.1, all contents X1 and meta data X2 are sent to the proxy in green (wide blue and red arrows); only the data of the intercepted subscribers are sent to the MF (the blue and red arrows become narrower).

To develop the "LI mediation system," this is a major task; one of the first things to do is to create an ASN1 coder/decoder according to [8.3]; it is used for the HI1 and HI2 interfaces.

There are a number of vendors: AQSACOM (FR, USA), BAE Systems (Datica, UK), Group 2000, Pine Digitas (NL), Verint (Israël), SS8 (USA), Suntech, and Ultimaco (DE) which provide this type of mediation system.

In many countries, the LEMF [8.4] is operated by the police under the control of the justice department. It accesses all the MNOs and Internet access providers.

8.2.2 LI management notification operation (HI1 interface)

It uses ASN1 encoded messages, while X1 [8.5] uses XML.

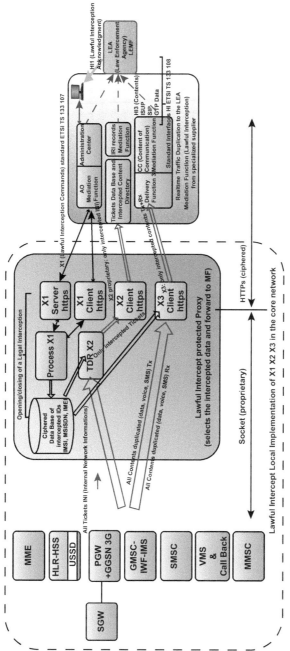

Figure 8.1 Mediation system to implement legal interception in an MNO, internal side and LEA side.

8.2.3 HI2: Handover interface port 2 (ASN1 coding)

Intercept-related information (IRI) shall be conveyed to the LEMF in messages, or IRI data records, respectively, using the HI2 interface (Handover Interface Port 2). As soon as the events are known, FTP may be used. Four types of IRI records are defined:

1. IRI-BEGIN record at the first event of a communication attempt, opening the IRI transaction.
2. IRI-END record at the end of a communication attempt, closing the IRI transaction.
3. IRI-CONTINUE record at any time during a communication attempt within the IRI transaction.
4. IRI-REPORT record used in general for non-communication-related events.

8.2.4 HI3: Handover interface port 3

This carries the CC records.

8.3 Services concerned by the interception need

This is a large list in [8.3] of 2017.

- Circuit switched domain (voice)
- Supplementary services (call forwarding, calling line number, etc.) (LI required in HLR-HSS)
- Packet data domain services (LI required only in the GGSN not in the SGSN)
- Multimedia domain
- 3GPP WLAN domain
- Multicast services (MBMS)
- Evolved packet system (LI only required in the PGW not in the MME)
- 3GPP IMS conference services
- 3GPP IMS-based voice services
- Proximity services (ProSe)
- Group communication service enables (GCSE)
- Messaging services (SMS and MMS).

8.4 Practical use of the content interception HI3 or X3

8.4.1 Use of meta data

One must understand that the full mass analysis of content is impractical as it would use an enormous number of police members, and also because some (data) are ciphered. For example, when data are captured, HTTPs cannot be deciphered because of the TLS security. This means that only the meta data such as the URLs and the protocol type can be extracted in all cases; they are sufficient to suspect criminal behaviors in some cases with automatic algorithms. Only HTTP can be displayed with software tools and a human analysis. The known deep packet inspection very sophisticated software objectives is only to extract these meta data, the IP assigned to the user for identification and pass them to a software which will make a history and decision.

For the voice conversations, they will be listened in a small number of cases; it would require too much police resources. On the other hand, the processing of the meta data can be automated; this is the aim as early detection of crimes and terrorism is essential.

8.4.2 Non-judiciary interception methods of security agencies

If the data content is strongly ciphered such as in the Telegram messaging, the content cannot be examined beyond knowing that it is a Telegram user. As this messaging is rather popular among terrorists or pre-terrorists, it is an important subject as the text messages can be automatically analyzed for the mass proactive terrorist detection. A known method of the security agencies is to insert spywares in the handsets of the targets, which can capture the passwords, etc. (remember the alleged US attack against the German chancellor). Inserting spywares is facilitated by the wide usage of non-secured OS in the handsets. If the potential targets access a frequently used website such as social security or income tax payment, or a GAFA application, the site could use a weakness of the OS to take over the control of the browser and then insert the spyware. Anyone which wishes to prevent to be infected by a spyware should use only secured phones with secured OS such a Secdroïd. Wise terrorists will also avoid accessing public official websites for the case they are connected with a national security agency (frequent in countries strong on their national security).

If this is the easiest solution, the security agencies could develop technical methods, with fake "IMSI catchers" stations to force discretely the data connection to their spyware infection site. This was easy with the obsolete 2G and the COMP128 algorithm [8.6], but in the case of Milenage 4G [4.13] with the forced roll-back to 3G being now protected, it is (2018) considered as unfeasible and hence the confidence in milenage. Installing infection mechanisms in the public site is certainly the most practical solution at a state level, of course, the major GAFAs can do it for years, and it is easy to defend the case if someone finds out and complains that his handset has been infected. The GAFA search engine accuses the Wikipedia website which was searched for the name of the scientist who resolved the Bale problem.[1]

---

[1]Elementary to resolve in fact, use the Taylor formula as it is called, expression of $\sin(x)$:

$$\sin(x) = x - \frac{x^3}{3!} + \frac{x^5}{5!} + \cdots$$

As $\sin(x)$ has zeros for $0, \pi, 2\pi, \ldots, k\pi, \ldots$, $\sin(x)$ must be of the form:

$$\sin(x) = K(x)\left(1 - \frac{x^2}{\pi^2}\right)\left(1 - \frac{x^2}{2^2\pi^2}\right)\left(1 - \frac{x^2}{3^2\pi^2}\right)\cdots$$

As $\lim \frac{\sin(x)}{x} \to 1$ when $x \to 0$, we have $K(x) = 1$, we have by dividing the two expressions of $\sin(x)$ by x:

$$1 - \frac{x^2}{3!} + \frac{x^4}{5!} + \cdots = \left(1 - \frac{x^2}{\pi^2}\right)\left(1 - \frac{x^2}{2^2\pi^2}\right)\left(1 - \frac{x^2}{3^2\pi^2}\right)\cdots$$

We have two expressions of $\sin(x)$ as left and right polynomials, the coefficients of x^2 in particular must be the same:

$-\frac{x^2}{3!} = -\frac{x^2}{\pi^2} - \frac{x^2}{2^2\pi^2} - \frac{x^2}{3^2\pi^2} \cdots$, then divide by $-x^2$ and multiply by π^2 to get Euler's result, see this chapter's quote.

References

[8.1] ETSI TS 133 106 V14.0.0 (2017-04), "Universal Mobile Telecommunications System (UMTS); LTE; 3G security; Lawful interception requirements" (3GPP TS 33.106 version 14.0.0, Release 14).

[8.2] ETSI TS 133 107 V14.1.0 (2017-04), "Universal Mobile Telecommunications System (UMTS); LTE; 3G security; Lawful interception architecture and functions" (3GPP TS 33.107 version 14.1.0, Release 14).

[8.3] ETSI TS 133 108 V14.0.0 (2017-04), "Universal Mobile Telecommunications System (UMTS); LTE; 3G security; Handover interface for Lawful Interception (LI)" (3GPP TS 33.108 version 14.0.0, Release 14).

[8.4] "La plateforme d'écoute de la justice agace la police," Le Figaro 25 October 2017, *article on the French national platform for lawful interceptions (PNIJ).*

[8.5] ETSI TS 103 221 V1.1.1 (2017-10), "Lawful Interception (LI) Part1; Internal Network Interface X1 for Lawful Interception." *HTTPS from the ADMF with XML payload.*

[8.6] Stuart Wray, "COMP128: A Birthday Surprise", swray@bournemo uth.ac.uk, a 2G algorithm which only needs 60000 trials to find the Ki key of a SIM card the use it for 3G or 4G in the Milenage algorithm.

9

Diameter-Based M2M (LTE-M and NB-IoT) 3GPP Services and LoRa

9.1 Operational need for M2M in PMR networks

A special architecture has been invented for LTE-M and IoT "indirect model" with new Diameter interfaces and terms. Was it really necessary to complicate, and which 4G/3G classical architecture would avoid these new MME or SGSN interfaces which are beyond the influence of the IoT service operators which use MNOs' radio coverage to connect the LoRa, Sigfox, or other technology 4G gateways? The architecture with the control plane below is applicable to the NB-IoT of [9.10] not to LoRa or SigFox which are pure end-to-end user plane architectures.

The emergence of LTE-M and NB-IOT requires a dedicated management and fully secured service in replacement of the traditional over the top (OTT) data service. MNOs can also deploy an additional dedicated cellular-IoT core.

9.2 3GPP NB-IOT type of services between the AS IoT server and the M2M devices: Direct, indirect, and hybrid model implementations

The AS may request periodically a long report. The AS may set an M2M device to send SMS when different events occur. SMSs are the basic mechanism for short messages which can a use a standard SMS service with the SGd interface in the MME or SGSN or even the E interface of the MSC/VLR. It is not obvious why a new protocol T6 was introduced as, in my opinion, an alternative to SGd, the lack of general implementation deprives it of any interest.

Different models are foreseen for machine type of traffic in what relates to the communication between the AS and the 3GPP network. The different architectural models that are supported by the architectural reference model are:

- Direct model – The AS connects directly to the operator network in order to perform direct *user plane* communications with the UE without the use of any external network signaling. The application in the external network may make use of services offered by the 3GPP system. Direct models are end-to-end and *do not concern this chapter*; the UE connects to the AS as a web service and remains connected in terms of session even if the bearer is released because of inactivity. The AS is connected by the Sgi/Gi interface to the PGW.
- Indirect model – The AS connects indirectly to the operator network through the services of a "gateway" service capability server (SCS) in order to utilize additional value-added services for MTC (e.g., *control plane* device triggering). The SCS is either:
- MTC service provider controlled: The SCS is an entity that may include value-added services for MTC, performing user plane and/or control plane communication with the UE.
- 3GPP network operator controlled: The SCS is a mobile operator entity that may include value-added services for MTC and performs user plane and/or control plane communication with the UE.
- Hybrid model: The AS uses the direct model and indirect models simultaneously in order to connect directly to the operator's network to perform direct user plane communications with the UE while also using an SCS. From the 3GPP network perspective, the direct user plane communication from the AS and any value-added control plane-related communications from the SCS are independent and have no correlation to each other even though they may be servicing the same MTC application hosted by the AS.

When using the hybrid model, the MTC service provider controlled SCS and the 3GPP operator controlled SCS may offer different capabilities to the MTC applications.

The TS 23.682 [9.1] facilitates communications with packet data networks and applications (e.g., machine type communication (MTC) applications on the (external) network/MTC servers) in replacement of traditional IoT architecture using a proprietary dialog between the device and the AS over a data channel 3G, 4G, or WiFi.

Since the different models are not mutually exclusive, but just complementary, it is possible for a 3GPP operator to combine them for different applications. This may include a combination of both MTC service provider and 3GPP network operator controlled SCSs communicating with the same PLMN.

Direct user plane methods are terminal dependent and may lack easy interworking, while 3GPP control plane methods are precisely defined and then adapted for a provider of an IoT service to third parties' AS allowing them to manage their worldwide mobile IoT devices such as containers and street lamps.

9.3 Necessary additional diameter protocols in an "IoT ready" EPC for the "indirect model"

It is natural to present the IoT support after the previous Chapter 5 on 4G SMS as the SMS delivery mechanism is a key item in the 4G IoT architecture. It allows the MTC-IWF with the T4/Diameter interface to trigger the SMS to send SMS to the IoT device.

So, an SMSC with an SM-IP-GW is part of what must be provided by the EPC.

Figure 9.1 is the standard architecture; the signaling between the MME and the HSS is not S6a but T6a with the SCEF which used the [9.9] introduced RestFull API protocol over T8.

There should be an OTA server using SMS or pull BIP/CAT-TP which is important for IoT, as it allows to add or change the IMSI security domain of the M2M SIM cards in the devices for a commercial change of their MNO. This is explained with a discussion of the card type UICC or eUICC in Chapter 4, where it is shown that the easily OTAble UICC meets the requirements for IoT.

9.3.1 T4 interface from the application IoT to the MTC-IWF and then to the SMSC

If the IoT application wants to request a device to perform an action such as returning a measurement or a report, it does it by sending an SMS. It submits the SMS to the MTC-IWF which uses the T4/Diameter to submit an SMS to the SMSC. T4/Diameter is a modern equivalent of the widely used SMPP V3.4 (Small Message Point-to-Point Protocol).

When messages are sent to the UE, the AVP User-Identifier contains the IMSI of the UE in general. It may contain the MSISDN if there is one and the SMSC knows it.

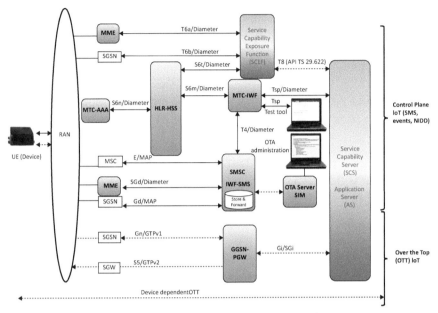

Figure 9.1   Architecture for LTE-M and IoT 3GPP from [10.1].

To complete accurately the general Figure 9.1, Table 9.1 gives the list of signaling messages between the MTC-IWF and the SMSC.

Table 9.1   T4 interface messages

T4 Command/Diameter TS 29.337	Short Name and Direction	Equivalent SMPP
Device Trigger Request	DTR SMSC ← MTC-IWF	submit_sm PID = 72 is used for the device triggering application, see [10.8]
Device Trigger Answer	DTA SMSC → MTC-IWF	submit_rsp
Delivery Report Request	DRR SMSC → MTC-IWF	deliver_sm
Delivery Report Answer	DRA SMSC ← MTC-IWF	deliver_rsp

1. The SCS determines the IP of the device by a Query DNS.
2. The SCS sends the Tsp Device Trigger Request to the MTC-IWF.
3. The MTC-IWF checks that the device is authorized with a Subscriber Information Request/S6m.

Table 9.2 Messages of the S6m and S6t interfaces

S6m, S6t Commands TS 29.336	Short Name and Direction	Parameter and Equivalence
Subscriber-Information-Request S6m	SIR HSS ← MTC-IWF	User-Identifier (IMSI UE, MSISN or external Identifier. *Equivalent to Update Location S6a*
Subscriber-Information-Answer S6m	SIA HSS → MTC-IWF	
Configuration-Information-Request	CIR HSS ← SCEF	User-Identifier (IMSI UE, MSISN or external Identifier. *Equivalent to Report SM delivery Status S6c for SMS*
Configuration-Information-Answer	CIA HSS → SCEF	
Reporting-Information-Request	RIR HSS → SCEF	User-Identifier (IMSI UE, MSISN or external Identifier. Monitoring Report (from the UE) *Equivalent to Alert-Service Center S6c for SMS*
Reporting-Information-Answer	RIA HSS ← SCEF	

Table 9.2 gives the list of signaling messages between the MTC-IWF and the HLR-HSS in Figure 9.1. Table 9.3 concerns the list of messages between an application and the MTC-IWF.

Figure 9.2 gives the details by including the Tsp, T4, and SGd messages. The DAR from the SCS can be considered as the SMS-MO sent to the MTC-IWF. The MTC-IWF does not work as an SMSC as it does not perform the Store&Forward if the UE is not reachable; this is done by the SMS 4G.

9.3.2 S6m and S6t interfaces with the HLR-HSS TS 29.336 [9.2]

The SGSN or MME communicates with the MTC-IWF when a UE object registers on the access network. It uses the T6b or T6a/Diameter interface. The MTC-IWF communicates with the HSS through S6m. The SCEF which we do not include as part of the core EPC because it also has proprietary interfaces on the SCS side (services capability server) and sets the HSS with CIR/S6t.

9.3.3 T6a and T6b interfaces between MME or SGSN and MTC-IWF TS 29.128 [9.4]

This is much like an SMS-MO and SMS-MT scheme between MME or SGSN and SCEF which may look superficially simpler than using the more general SGd interface of MME or SGSN which also works for ordinary UEs. If the core network has a full SMSC with IP-SM-GW, there is no reason to base the IoT service on this interface.

9.3.4 Tsp interface between the SCS and the MTC-IWF (TS 29.368 [9.7])

Table 9.3   Messages of the Tsp interface

Tsp Commands TS 29.368	Short Name and Direction	Parameter and Equivalence
Device-Action Request	DAR MTC-IWF ← SCS	Device Action
Device-Action Answer	DAA MTC-IWF → SCS	
Device-Notification Request	DNR MTC-IWF → SCS	Device Notification
Device-Notification Answer	DNA MTC-IWF ← SCS	

9.3.5 Implementation strategy, what is the simplest and most general implementation for IoT

To have a general-purpose IoT core, it is logical to include the IoT-capable HLR-HSS, the SMSC 4G, and the MTC-IWF with all their Diameter interfaces, that is, the functions included in the trust IoT domain. An SCEF and an SCS-AS must be added to complete the system, using the newly (2017) standardized T8 API [9.10] between the SCEF and the SCS-AS.

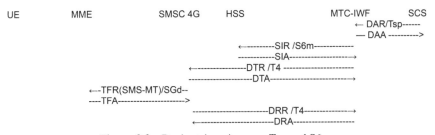

Figure 9.2   Device triggering over Tsp and S6m.

9.4 LoRa

LoRa [9.5] is a low-consumption and simple technology compared to 3G which has a significant deployment with many major operators having created LoRa networks. We will particularly discuss the implementation of roaming to allow devices of a mobile fleet to be able to use several LoRa infrastructures with the possibility of automatic handover in Rel 1.1 only. LoRa coverage is not expensive (12,000 euros only in 2017 to cover the Roissy Charles de Gaulle airport with seven LoRa antennas and gateways installed on existing masts).

9.4.1 The LoRaWAN architecture

The LoRa architecture is explained by comparing to the assumed known 3G architecture; it is extremely simplified which makes a rather simple project to develop all the network elements.

The explanations are given in order to explain a LoRa roaming hub using passive roaming as this is applicable to the current LoRaWAN R1.0 while handover roaming is only for R1.1 [9.10] devices, so it extends the scope.

9.4.1.1 End-Device (equivalent of combined UE + SIM Card)

The end-device is a sensor or an actuator such as a camera or a temperature probe. The end-device is wirelessly connected to a LoRaWAN network through LoRa radio gateways. The application layer of the end-device is connected to a specific AS in the cloud. All application layer payloads of this end-device are routed to its corresponding AS. The Join Request message contains a NONCE, equivalent of the Sequence Number in a SIM card to protect against replay.

9.4.1.2 Radio gateway (equivalent of an RNC 3G)

The radio gateway forwards all received LoRaWAN radio packets to the network server (NS) that is connected through an IP back-bone. The radio gateway operates entirely at the physical layer. Its role is simply to decode uplink radio packets from the air and forward them unprocessed to the NS. Conversely, for downlinks, the radio gateway simply executes transmission requests coming from the NS without any interpretation of the payload.

9.4.1.3 Network server (equivalent of SGSN 3G (no GGSN equivalent used))

The NS terminates the LoRaWAN MAC layer for the end-devices connected to the network (the same role as an MME). It performs the DNS resolution based on the network ID (equivalent of MCC-MN) in the DevEUI to find the JoinServer (authentication role of an HSS) and then perform the routing of the application packets to the AS directly. There is no GGSN equivalent need because the user data are not encapsulated in GTP-U and there is no equivalent of APNs with the possibility of different QoS and dynamic controls. The LoRa generic features of NS are:

- End-device address check;
- Frame authentication and frame counter checks;
- Acknowledgments;
- Data rate adaptation;
- Responding to all MAC layer requests coming from the end-device;
- Forwarding uplink application payloads to the appropriate ASs;
- Queuing of downlink payloads coming from any AS to any end-device connected to the network;
- Forwarding join-request and join-accept messages between the end-devices and the join servers (JSs).
- In a roaming architecture, an NS may play three different roles depending on whether the end-device is in roaming situation or not, and the type of roaming that is involved,
- Serving NS (sNS) controls the MAC layer of the end-device,
- Home NS (hNS) is where device profile, service profile, routing profile, and DevEUI of the end-device are stored. hNS has a direct relation with the JS that will be used for the join procedure. It is connected to the AS. When hNS and sNS are separated, they are in a roaming agreement. Uplink and downlink packets are forwarded between the sNS and the hNS.
- Forwarding NS (fNS) is the NS managing the radio gateways. When sNS and fNS are separated, they are in a roaming agreement. There may be one or more fNS serving the end-device. Uplink and downlink packets are forwarded between the fNS and the sNS.

The full list of messages of the LoRa "backend protocol" is given in Table 9.4.

Table 9.4   "Back end" protocol [9.11] message types

Request	Answer
JoinReq	JoinAns
RejoinReq	RejoinAns
AppSKeyReq	AppSKeyAns
PRStartReq	PRStartAns
PRStopReq	PRStopAns
HRStartReq	HRStartAns
HRStopReq	HRStopReq
HomeNSReq	HomeNSAns
ProfileReq	ProfileAns
XmitDataReq	XmitDataAns

9.4.1.4 Central DNS of the LoRa alliance

The NS makes a DNS resolution to get the IP of the JS and also the IP of the hNS in Figure 9.3. The LoRa alliance establishes and operates two dedicated subdomains to reach the JS and NetIDs, rooted at JOINEUIS.LORA-ALLIANCE.ORG and NETIDS.LORAALLIANCE.ORG, respectively.

1) A Join EUI (IEEE EUI-64) is represented as a name in the JOINEUIS.LORA-ALLIANCE.ORG domain by a sequence of nibbles separated by dots with the suffix ".JOINEUIS.LORA-ALLIANCE.ORG." The sequence of nibbles is encoded in reverse order, i.e., the low-order nibble is encoded first, followed by the next low-order nibble, and so on. Each nibble is represented by a hexadecimal digit. For example, the domain name corresponding to the EUI:

$$00\text{-}00\text{-}5E\text{-}10\text{-}00\text{-}00\text{-}00\text{-}2F$$

would be:

f.2.0.0.0.0.0.0.0.1.e.5.0.0.0.0.JOINEUIS.LORA-ALLIANCE.ORG

The NAPTRs point to replacement servers according to order, preference, and flags, and service parameters provided by the operators. In Figure 9.3, 1) the LORA DNS will return the IP and port of the JS 00-00-5E-10-00-00-00-2F.

2) Later, in Figure 9.3 12), the AS is resolved by a DNS request using the Home-NetID in the JoinAns. A 24-bit NetID is represented as a name in the NETIDS.LORA-ALLIANCE.ORG domain by a sequence of nibbles with the suffix ".NETIDS.LORA-ALLIANCE.ORG." The high-order nibble is encoded first, followed by the next higher order nibble, and so on. Each nibble is represented by a hexadecimal digit. For example, the domain name

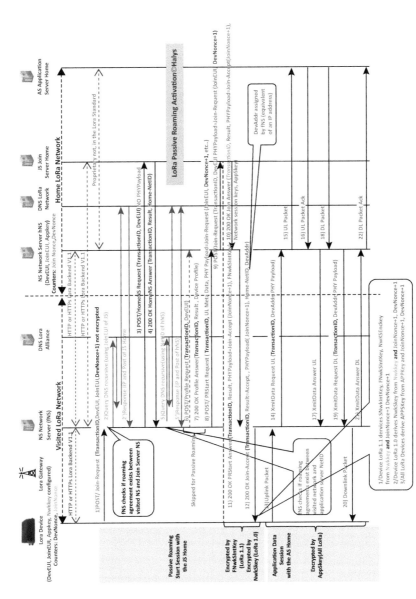

Figure 9.3   LoRaWAN architecture and passive roaming activation procedure.

corresponding to the NetID which has been returned in the Join-Answer message:

$$1290 \ (0x00050A)$$

would be

$$00050a.NETIDS.LORA-ALLIANCE.ORG$$

and it is used by the sNS to find the IP of the AS of the device.

Table 9.5 characterizes the identities used in the LoRa protocol.

Table 9.5 Join-request message network server→join server

JoinEUI	DevEUI	DevNonce
8 octets (used to *resolve the home Join Server*)	8 octets assigned by manufacturer (equivalent of MAC address)	2 octets

9.4.1.5 Join server (equivalent of an HLR for just the authentication function)

The join procedure is always initiated from the end-device by sending a join-request message to the hNS which is the equivalent of a SEND AUTHENTI-CATION. The visited sNS resolves the JoinEUI with the LoraAlliance DNS to get the IP of the JS and sends it the Join Request message.

The JS returns the NetID of the device's home LoRa network as in Table 9.6.

The join-request message contains the JoinEUI and DevEUI of the end-device followed by a nonce of 2 octets (DevNonce). DevNonce is a counter

Table 9.6 Join-accept message network server → join server

JoinNonce	Home-NetID	DevAddr Device address not unique		DL setting	RxDelay	CF List
3 octets	3 octets (used to *resolve the home Network Server*) hNS	4 octets		1	1	16
		7 bits assigned to network 0x06=La poste 0x09=KPN, 0x0F=Orange France. etc.	25 bits			

starting at 0 when the device is initially powered up and incremented with every join-request. A DevNonce value SHALL NEVER be reused for a given JoinEUI value. If the end-device can be power-cycled, then DevNonce SHALL be persistent (stored in a non-volatile memory). Resetting DevNonce without changing JoinEUI will cause the NS to discard the join-requests of the device. For each end-device, the NS keeps track of the last DevNonce value used by the end-device and ignores Join-requests if DevNonce is not incremented.

This mechanism is designed to prevent replay attacks by sending previously recorded join-request messages with the intention of disconnecting the respective end-device from the network. To protect the radio transmission being intercepted, it is encrypted with the device's root keys NwkKey and AppKey which are shared in the device and the JS.

The JS manages the OTA end-device activation process. There may be several JSs connected to an NS, and a JS may connect to several NSs.

The end-device signals which JS should be interrogated through the JoinEUI field of the join-request message. Each JS is identified by a unique joinEUI value. Note that AppEUI field of the join-request in LoRaWAN 1.0/1.0.2 is renamed to JoinEUI field in LoRaWAN 1.1 [9.11]. The term JoinEUI is used to refer to AppEUI in the context of LoRaWAN 1.0/1.0.2 end-devices in the older specification.

The JS knows the end-device's hNS identifier and provides that information to the other NSs when required by the roaming procedures.

The JS contains the required information to process uplink join-request frames and generate the downlink join-accept frames. It also performs the network and application session key derivations. It communicates the network session key of the end-device to the NS and the application session key to the corresponding AS.

For that purpose, the JS SHALL contain the following information for each end-device under its control:

- DevEU;
- AppK;
- NwkKey (only applicable to LoRaWAN 1.1 end-device);
- hNS identifier;
- AS identifier;
- A way to select the preferred network in case several networks can serve the end-device;
- LoRaWAN version of the end-device (LoRaWAN 1.0, 1.0.2, or 1.1).

The root keys NwkKey and AppKey are available only in the JS and the end-device, and they are never sent to the NS nor the AS.

Secure provisioning, storage, and usage of root keys NwkKey and App-Key on the end-device and the backend are intrinsic to the overall security of the solution. These are left to implementation and out of scope of this document. However, elements of this solution may include SE (secure elements) and HSM (hardware security modules).

The way this information is actually programmed into the JS which is outside the scope of this document and may vary from one JS to another. This may be through a web portal, for example, or via a set of APIs.

The JS and the NS SHALL be able to establish secure communication which provides end-point authentication, integrity, replay protection, and confidentiality. The JS SHALL also be able to securely deliver the application session key to the AS.

The JS may be connected to several ASs, and an AS may be connected to several JSs.

The JS and the AS SHALL be able to establish secure communication which provides end-point authentication, integrity, replay protection, and confidentiality.

9.4.1.6 Application server

The AS handles all the application layer payloads of the associated end-devices and provides the application-level service to the end-user. It also generates all the application layer downlink payloads toward the connected end-devices.

There may be multiple ASs connected to an NS, and an AS may be connected to several NSs (operating end-devices through several networks, for example). An AS may also be connected to multiple JSs.

The hNS routes the uplinks toward the appropriate AS based on the DevEUI.

In addition to the aforementioned network elements, LoRaWAN architecture defines the following network interfaces among these entities:

hNS-JS: This interface is used for supporting the join (activation) procedure between the JS and the NS.

vNS-JS: This interface is used for roaming activation procedure. It is used to retrieve the NetID of the hNS associated with the end-device.

ED-NS: This interface is used for supporting LoRaWAN MAC-layer signaling and payload delivery between the end-device and the NS.

AS-hNS: This interface is used for supporting delivery of application payload and also the associated meta data between the AS and the NS.

hNS-sNS: This interface is used for supporting roaming signaling and payload delivery between the hNS and the sNS.

sNS-fNS: This interface is used for supporting roaming signaling and payload delivery between the sNS and the fNS.

AS-JS: This interface is used for delivering the application session key from the JS to the AS.

In Figure 9.4:

The connection is periodical depending on the device class (A = rarely autonomy can be several years, C = almost always connected).

The session is established through the visited NS (fNS). It must find the IP of the JS. It interrogates (2 and 2') recursively the Home DNS starting with the root DNS of the LoRa alliance. It gets the IP of the JS and sends it a HomeNS Start (3), with an answer (4) containing the identity Home-NetID of the hNS which serves the AS.

To establish the connection with the hNS, the fNS makes a second recursive DNS query (5 and 5') using Home-NetID as argument and it gets the IP of the hNS.

Steps 6) and 7) are not used in passive roaming.

In 8), the fNS sends a PRStart Request to the hNS, containing DevNonce+1 in the PHYPayload. Until this step 8), the payload is not encrypted.

In 9), the hNS sends a request join-request to the home JS, which contains the same PHYPayload as in the PSStart request (8). The JS may thus check that DevNonce is greater than 1 above the last received for the device. If yes, it answers join-answer (success) with its own sequence JoinNonce + 1.

The hNS (11) responds a PRStart answer "success" to the interrogating fNS; it contains JoinNonce+1 then. From then, it is symmetrically encrypted with NwkSKey that the hNS has computed from NwkKey known from the device and the new DevNonce and JoinNonce.

The fNS answers (12) to the fNS by a Join-Accept which contains Join-Nonce+1 as well as DevAddr (equivalent of a dynamic IP address) which is used for the bidirectional transmission. The device receives it, knows its own DevNonce+1, and is able to assume the JoinNonce+1; it can also compute the same NwkSKey and use it to decipher the Join-Accept. If the JoinNonce +1V is what is expected, it establishes the session.

There are TransacTion IDs to allow parallel sessions.

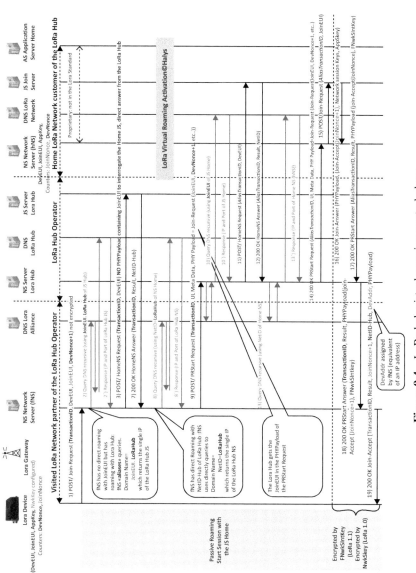

Figure 9.4 LoRa hub virtual roaming control plane.

9.4.1.7 Packet transmission

The application data session in Figure 9.3 illustrates the message flow for passive roaming procedure 515354091 between the fNS and the hNS two NSs. The packets are encrypted with the AppSKey simultaneously computed by the device and the hNS.

The AS is connected to the hNS but the specification is not included in LoRa. It could be http or https with POST or GET queries or some proprietary protocol decoded by the Home LoRa network.

9.4.2 Device classes

- A: Can receive a message only at the time the device is sending a message.
- B: The same as A but listens for incoming messages on regular intervals.
- C: Continuously listening for incoming messages.

Logically, class A devices can last the longest on battery and class C devices are typically not battery powered.

The LoRaWAN standard describes two different types of messages: MAC messages, to control the radio and network and data messages, the actual payload that is application/device specific. Since we are limited in the time and number of messages we can send over the air, MAC messages can be piggybacked (send together) with data messages and multiple MAC messages can be sent in one time.

9.4.3 Device addressing, LoRa roaming, and LoRa hubs

As with most networking standards, devices need some kind of address and identification to be able to contact them and differentiate them from each other. LoRa uses the following addressing.

- DevEUI: Device unique hardware ID: 64 bits address. Comparable with a MAC-address for a TCP/IP device.
- DevAddr: Device address: 32 bits address assigned or chosen specific on the network. Comparable with an IP address for a TCP/IP device.
- JoinEUI: JS address: EUI64 address format. Uniquely identifies the JS of the device. Then JoinEUI is stored in the end-device before the activation procedure is executed.
- Fport: It identifies end application/service. Port 0 is reserved for MAC messages. Comparable with a TCP/UDP port number for a TCP/IP device.

Technically, the LoRa gateways can address several NSs based on the application ID if an agreement exists. This means that a LoRa hub could aggregate agreements with many LoRa networks and relay the uplink traffic to several NSs. It could also be directly the NS for direct clients with their AS.

9.5 LoRa virtual roaming hubs

LoRa roaming faces a major difficulty because different LoRa frequencies are used in the world. *LoRa does not have a "beacon channel"* and the devices cannot passively know which frequency to use to transmit without knowing which country they visit. To generalize LoRA roaming in various countries without the same standard, devices with a low-power GPS or a GSM receiver listening the Beacon channels will need to be developed; there is a market with the ship containers. Assume (Europe for example) that this is not an issue (868 band [9.12]).

9.5.1 Architecture principles

A LoRa hub uses some of the same principles as a GTP hub [0.9]. It behaves as *the home JS and the hNS for the visited LoRa fNS* and as *a fNS for the home JS and the hNS*. It is explained with the help of the detailed Figure 9.4.

The method of the invention [9.13] includes a way to resolve the main difficulty: the DNS resolution (7) before sending the HomeNS request (8) needs as argument the JoinEUI of the Home JS: the LoRa Hub extracts this JoinEUI it from the PHYPayload (containing the Join-Request message 1), itself containing the needed JoinEUI.

Another way is to use the virtual IP address idea (one distinct IP address belonging to its own range is assigned by the LoRa Hub for each customer LoRa network). This IP is returned 2') in the recursive DNS query sent 2) to the LoRa alliance root DNS. When the LoRa hub receives the HomeNS request 3), it recognizes the IP and can find the JoinEUI with a table.

$$\text{Destination IP of the JS} \rightarrow \text{JoinEUI}$$

This second way is reserved for LoRa hub carriers which have available a large number of IPv4 addresses; it is given for the reader's understanding, as the first way applies to all cases and requires a single IP address on the Internet.

The LoRa protocol also includes Transaction ID which allows different simultaneous sessions; they are already assigned when the fNS sends the HomeNS request (3) to the Lora hub JS. When the messages are relayed to the Home LoRa network, "aliasing" must be used in order to avoid a possible duplication of the Transaction IDs. This is the same principle as with GTP Hubs which must "alias" the TEID (terminal end-point identifier).

Another principle of the invention is that *the LoRa Hub is not involved in the end-to-end encryption security between the Device and the hNS*. It relays transparently the end-to-end DeviceNonce and JoinNonce as well as NetID, so it needs no knowledge of the NwkKey and AppKey. The User Data uplink and downlink pass through the LoRa Hub so it can provide its charging role and the clearing if the LoRa Virtual roaming agreement is symmetrical.

9.5.2 Setups between a LoRa hub operator and visited LoRa network partners

The Lora hub operator, e.g., LoraHubFR, will start by making a number of commercial agreements with LoRa networks, those which deploy LoRa gateways and at least one NS which will be the fNS for the visiting virtual roaming devices, for example, in Europe Orange and Objenious (France), Proximus (Belgium), Digimondo (Germany), in the US Sinet, etc., and LoRa NSs in various countries.

The agreement has a technical part: the visited network LoRa gateway, instead of rejecting as in Figure 9.3 (2) a request for a JoinEUI, it does not have a roaming agreement with, will "alias" the DNS request with a domain name (for example, LoraHubFR) which in the DNS recursion interrogates the LoRa hub operator:

f.2.0.0.0.0.0.0.0.1.e.5.0.0.0.0.LoraHubFR.JOINEUIS.LORA-
ALLIANCE.ORG

As explained above, this scheme allows the LoRa hub to *know which JoinEUI is concerned* when it receives the hNS request which does not contain it.

The commercial part of the agreement is that the LoRa network sends all the charging records to the used LoRa hub operator which pays them.

9.5.3 Detailed explanation of the LoRa hub operation

9.5.3.1 Commercial aims: No agreement between the visited and home LoRa networks, one-stop shopping with the LoRa hub operator

The many end-user ASs which need LoRa mobility may directly contact (a tariff option) with a LoRa operator in their country which is backed by an agreement with the LoRa hub. In this case, there is no direct agreement between the LoRa hub operator and the many ASs which retain their connection and contract with a national LoRa network, connecting to their NS (hNS in Figure 9.4). The national LoRa network is charged by the LoRa hub operator which charges its AS customer.

To allow the use of a LoRa Hub, the fNS has created an "aliasing" in its NS: all requests concerning a visiting device which network (JoinEUI does not have an agreement with is sent to the LoRa Hub). Aliasing means that the DNS argument to find the JS becomes JoinEUI.LoRaHub. LoRaHub is the JoinEUI which the fNS has the agreement with.

9.5.3.2 Explanation of the control plane call flow

Let us add some details in the explanations included in Figure 9.4. The connection is the first part.

- 1) The device sends a join-request with its JoinEUI to the visited fNS.
- 2) and 2') The fNS recognizes that it does not have a roaming agreement and aliases the control plane LoRa messages, so they are sent to the lora roaming hub.
- The HomeNS Request sent by the fNS 3) is responded directly 4) by the LoRa Hub JS which returns its own NetID = LoRaHub.
- 5) and 5') The fNS recognizes that it has a roaming agreement with LoRaHub and sends 6) the PRStart request to the LoRa hub NS.
- 7) and 7') The LoRa hub NS, as explained above, queries the DNS root recursively with JoinEUI to obtain the IP of the Home JS.
- 8) The LoRa hub NS sends a HomeNS request to the Home JS containing the DeviceEUI. It is answered 9) with the Home NetID.
- 10) and 10') The LoRa hub NS queries the DNS with this NetID to get the IP and Port of the hNS of its customer.
- 11) The LoRa hub NS sends a PRStart request to the hNS.
- 12) The reception of PRStart answer is followed 12) by a PRStart answer back to the fNS 13), note the "de-aliasing" of the transaction ID.
- 14) The fNS responds a join-accept, for which it has assigned a DeviceAddr.

9.5.3.3 User data call flow for a LoRa hub

Once the LoRa session is established, the device and the hNS which both

- have the AppKey and NwkKey of the subscribing device;
- receive the JoinNonce and DeviceNonce, respectively;

are able to compute the same AppSKey used for the ciphering of the user data as shown in Figure 9.5.

In the control plane call flow, Figure 9.4 is very comprehensive for the LoRa session activation. Using [9.11], it is left as an exercise to the reader to make a call flow of the LoRa session deactivation.

9.5.4 Geo-localization in LoRa and applications: TDOA is the most appropriate method

The GPS-based geo-localization methods which require a GPS circuit in the terminal would draw quickly all the battery of a LoRa device; for applications such as container tracking and baggage tracking, they are not usable. The most appropriate method is time difference of arrival (TDOA) where a LoRa device sends a reference message containing a time Tue *which is not GPS synchronized*. The signal is received by *several GPS synchronized LoRa gateways* i, j, k,..n (e.g., in an airport) which are able to compute together the $n(n-1)/2$ TDOAs. The accuracy of each time measurement by the LoRA gateways is 30 ns ($\sigma = 90$ m on a distance measurement). There must be in the LoRa network some SMLC function which receives the n measurements simultaneously:

Ti – Tue time differences of the various LoRa gateways

The SMLC can then compute the $n(n-1)/2$ pairs

$$Ti - Tue - (Tj - Tue) = c\ (di- dj), c = 300,000 \text{ km/s}$$

which defines hyperbola with foci F_i at the various LoRa gateways (a hyperbola is the set of all points with a given difference of distance di – dj to two focal points Fi and Fj, a well-known elementary geometry result). The SMLC computes the position closest to all of them (point **P** in Figure 9.6). Below is the case for three LoRa gateways which send their measurements to the NS which has an SMLC using the same algorithm as in 3G or 4G U-TDOA.

The devices may use a low periodicity of a few hours (sufficient for baggage or container tracking) to preserve the battery and send a proprietary message containing their Tue which is decoded *by several different LoRa*

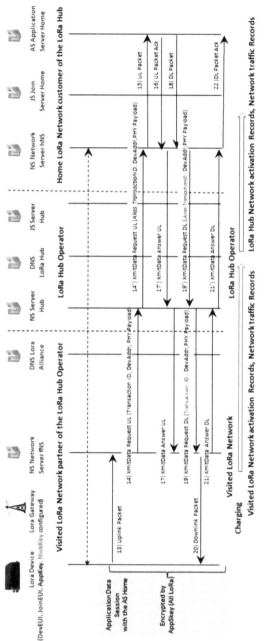

Figure 9.5 LoRa hub virtual roaming user data flow.

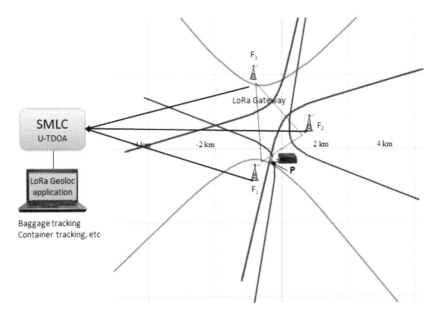

Figure 9.6   LoRa hyperbolic geo-localization with U-TDOA and an SMLC LoRa.

gateways Ti which receive it. These LoRa gateways send a message containing Ti – Tue from their Source IP#i address (they are connected by WiFi, 3G, or 4G and hence have an IP address) to their SMLC which is able from a table IP#i → Longitude–Latitude to determine the focal points Fi of all the hyperbolas.

Hence, an estimate of the device's location is found by computing the intersection of $n(n-1)/2$ hyperbolas as thoroughly explained with the mathematics in [0.8, Chapter 6 for E-OTD] or [0.10, Section 2.12.3.4 for U-TDOA] (in TDOA, the message is sent by the device and in E-OTD by the radio stations). The method does not use "triangulation", a non-liking geometry journalist's term suggesting that the position is obtained by intersecting straight lines from the radio station (this is the angle of arrival method which hardly exists even in recent 4G stations), but more exactly "hyperbolic intersections". With the device visible from three LoRa gateways in an airport, the accuracy would be about 90/sqrt (3) = 52 m, quite sufficient for a baggage tracking application; its real usefulness depends on whether the police finds the lost luggage abandoned before the airline and blows it up according to their stern duty.

The method for LoRA makes one think, because of the name similarity, of the old LORA aircraft navigation system of WWII which also used the intersection of hyperbolas, corresponding to a given phase shift of the low-frequency radio signals of two powerful land-based stations, nothing new then.

References

[9.1] 3GPP TS 23.682 v15.0.0 (2017–03), "Digital cellular telecommunications system (Phase 2+) (GSM); Universal Mobile Telecommunications System (UMTS); LTE; Architecture enhance10..1] tents to facilitate communications with packet data networks and applications," Release 13. *IoT standard architecture.*

[9.2] 3GPP TS 29.336 v14.1.0 (2017–05), "Universal Mobile Telecommunications System (UMTS); LTE; Home Subscriber Server (HSS) diameter interfaces for interworking with packet data networks and applications," Release 13, *The S6m/S6n/S6t protocols Diameter interfaces with a HSS "ready IoT."*

[9.3] 3GPP TS 29.229 v14.2.0 (2017–03), "Universal Mobile Telecommunications System (UMTS); LTE; Diameter-based T4 Interface for communications with packet data networks and applications," Release 14, *Protocol Diameter on interface T4 between MTC-IWF and SMSC.*

[9.4] 3GPP TS 29.128 v14.2.9 (2017–03), "LTE; Universal Mobile Telecommunications System (UMTS); Mobility Management Entity (MME) and Serving GPRS Support Node (SGSN) interfaces for interworking with packet data networks and applications," Release 14, *Protocol Diameter on interface T6a, T6b between MME or SGSN and MTC-IWF of the IoT ready core. To be IoT ready, the MMEs and SGSNs must have this interface as well as SGd to have the SMS capability of course.*

[9.5] 3GPP Low Power Wide Area Technologies, GSMA (PDF).

[9.6] Gsma, Official document CLP.28, NB-IoT Deployment Guide to Basic Feature Set Requirements, 2 August 2017.

[9.7] 3GPP TS 29.368 V14.2.0 (2017–07), "Universal Mobile Telecommunications System (UMTS); LTE; Tsp interface protocol between the MTC Interworking Function (MTC-IWF) and Service Capability Server (SCS)," Release 14. *Tsp/Diameter between the SCS and the MTC-IWF.*

[9.8] 3GPP TS 23.040 V14.0.0 (2017–04), "Digital cellular telecommunications system (Phase 2+) (GSM); Universal Mobile Telecommunications System (UMTS); Technical realization of the Short Message Service (SMS)," Release 14. *This release includes some parameters for the IoT SMS.*

[9.9] 3GPP TS 29.122 V0.2.0 (2017–09), "T8 Reference point for Northbound, bewared APIs," Release 15, *uses the RESTfull protocol.*

[9.10] LoRa Alliance, "LoRaWAN 1.1 specification," October 2017.

[9.11] LoRaWAN, "Backend interfaces 1.0 specification," Oct 11 2017, *description of the passive and handover roaming architecture, the document which explains roaming, beware there are errors in Table 13. Figure 9.4 of this book is more correct.*

[9.12] LoRaWAN, "Regional Parameters 1.1," 2017, *includes the various frequencies assigned to LoRa with "Common Name": EU868, US915, CN779 (China), IN865 (India), RU864 (Russia), KR920 (Korea), about 10 in all in various countries.*

[9.13] S. Cruaux, A. Henry-Labordère, B. Mathian, "Système de communication entre des objets connectès, en itinèrance et leur réseau," Patent FR 1852631, March 2018.

10

Advanced Policy Control and Charging (PCC), Standard Provisioning Architecture for HLR-HSS and PCRF

Without Leibnitz, Newton, d'Alembert, Euler, Wallis, and Gauss there would not be much matter teachable in maths for first and second year university, in third year nothing without Fourier and Cauchy. Anything else is for graduate studies.
Math teacher

10.1 Destination IP-dependent charging with an external DPI (deep packet inspection)

One wants to provide the charging of the data service depending on the web destination IP if international destinations are charged at a higher rate than national destinations. This requires a special equipment "DPI" able to analyze the Gi/SGi UL flow of up to tens of GB/s with a large table of IP addresses to recognize the national destinations and also PORTS such as those used for Skype.

Architecture

To do this, there is an external DPI (deep packet equipment) equipment which performs the charging using Gy/Diameter (TS 32.299) and sets the charging rate with the following AVPs set in the CCR that it sends to the online charging system (OCS):

- Rating-group AVP
- Multiple-services-credit-control

The PGW is *not involved in the charging* and performs only the *initial policy control* (with the PCRF) of the QoS. It also implements the creation of modification of the dedicated bearers when it receives RAR/Diameter from the PCRF as requested by a telephony application server (TAS) using its Rx/Diameter with the PCRF.

Concerning the IP destination charging, in an HTTP session, there may be many varying destination IPs and the DPII must dynamically send new CCRs/Gy to the OCS with a change of the rating-group AVP.

The equipment involved in the policy charging and control pcc are then:

- the GGSN-PGW for control
- the PCRF for control
- the DPI equipment for charging

RADIUS Interface PGW → DPI

To communicate charging identities, IP source, MSISDN, IMSI, APN, the PGW and the DPI communicate with the RADIUS protocol using Accounting-Request sent by the PGW and Accounting-Response answered by the DPI which behaves as a RADIUS server. The DPI is the able with the IP source to access the UE's charging identities for the Gy.

Also the PGW includes the Session ID (PCRF) to be used in the CCR/Gx after the RADIUS request.

10.1.1 Architecture consequence to satisfy the rerouting requirement

When there is no credit (recognized by the DPI), the session is interrupted. The UE will automatically reconnect and must then be rerouted to the credit recharging function. This means that the Destination IP must be dynamically changed by the DPI which entails that it is not just monitoring the Gi/SGi UL traffic. Hence, the *DPI must be inserted in the Gi/SGi flow* as illustrated in Figure 10.4.

10.1.2 Rating plan-dependent charging

The changing of voice services is customer dependent depending on his rating plan defined in the HLR-HSS which forwards specific service keys in the ISD response to the UPDATE LOCATION. There is no such thing for charging as Camel is not generally available for data roaming. In response to the UPDATE LOCATION GPRS, the only UE specific parameter is the MAP "Charging Characteristics." This parameter is operator specific and the same in general for all UEs.

Also a PCRF interrogation would not give the rating information. The "charging rules" of Gx (TS 29.212) concern only the TFT.

This means that if the operator wants to apply *rating plan-dependent charging* for the UEs, this must be implemented in the OCS which *has to be provisioned with individual subscriber profiles* giving the data charging plan or one must use the Charging Characteristics parameter of the HLR-HSS.

10.1.3 Data charging diagram

The DPI behaves as a radius server for the GGSN-PGW answering the Accounting-Request (Start–Stop); there is no Interim-Update It uses then to create or suppress a session context accessed by the IP Source address for the UL direction according to Figure 10.1.

Figure 10.2 illustrates a session disconnected UE is initiated and an Accounting-Request Stop is sent to the DPI which sends a CCR (Stop) to the OCS.

Figure 10.3 illustrates a session disconnection by the lack of credit recognized by the DPI.

10.1.4 Slow-down policing with an external DPI

In the above design, the QoS is entirely managed by the PGW, the PCRF, and some AS which is Rx capable according to the 3GPP standards. Could a *"slow-down policy"* be developed? When it recognizes (IP or PORTS) that a service must be slowed down, the DPI would dynamically change the QoS by either:

$\rightarrow$ Solution 1 (Figure 10.4): Behaving as a PCRF, which sends RAR (Re-Auth Request)/Gx Diameter to the PCEF of the PGW to reduce the rate. This is the clean method without packet dropping as the enforcement of the policy is performed by the RAN. With this option, the DPI equipment would also be involved in the control of the QoS.

$\rightarrow$ Solution 2 (Figure 10.5): Behaving as a session border controller, it needs to have the Rx/Diameter protocol.

Whatever is the solution this is an option which is not needed for just the flexible charging objective national/international. But it is useful to implement the slowing down of Facebook, Uber, Google, etc. All the GAFAs push for the states' investments in large band while making the profit. The solution presented allows implementing a non-neutral "Internet policy."

This is a war between the "defenders of liberty" and the countries which historically defend the privacy of their citizens and national security. The fight between the contradictory RGPD law and the countering US Cloud Act [10.5] may be finally a good thing, as any consumer of a UE bank, insurance, public

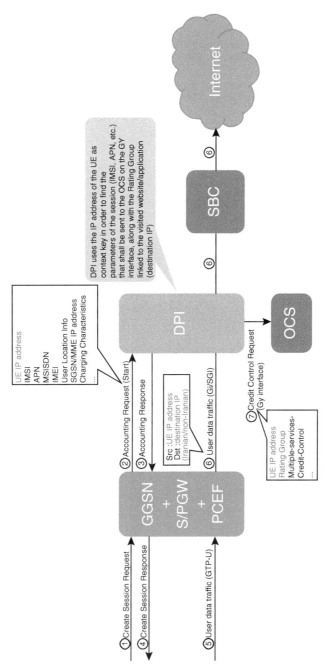

Figure 10.1   Context creation in the DPI and destination-based charging with OCS.

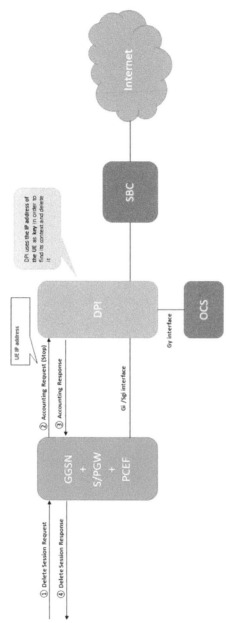

Figure 10.2 Context delete in the DPI.

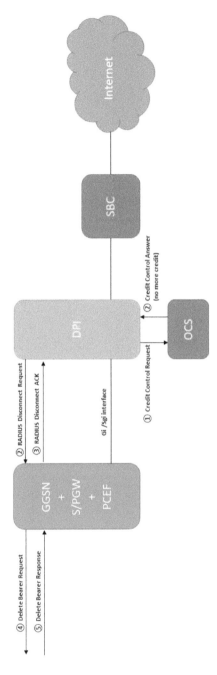

Figure 10.3   Session deletion triggered by no more credit indication.

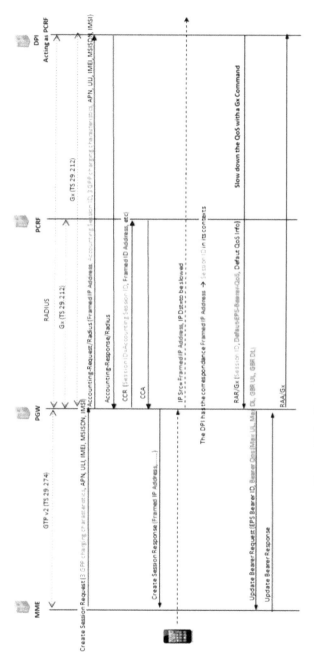

Figure 10.4   Slow-down policing option (Gx/Diameter based).

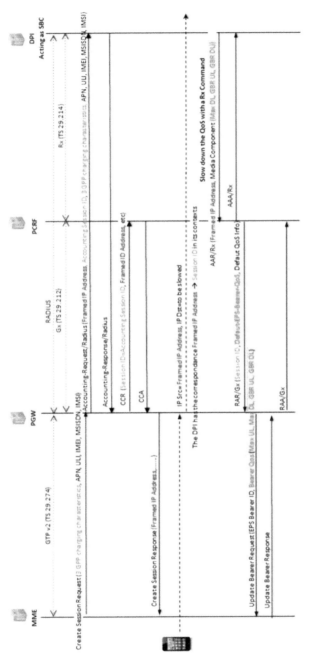

Figure 10.5   Slow-down policing option (Rx/Diameter based).

service may require that no access may be possible from any GAFA service as it would mean that his personal data may be transferred to the US under their Cloud Act US obligation, this being contradictory with the RGPD UE law. If legal actions are taken, this could me the barring of many GAFA services.

10.1.4.1 Gx-based slow-down policing implementation in the DPI

The Session ID (PCRF) included in the DPI context allows the DPI which has the Framed IP address to send a RAR with the Session ID to the PGW.

The DPI may recognize certain destinations and would like to apply a "slow-down policy". It would send a RAR/Gx to the PCEF in the PGW.

In order to do this, the DPI would use the *Session ID* (for the PCRF) which is the same as the RADIUS Acct-Session-ID.

10.1.4.2 Rx-based slow-down policing implementation in the DPI

The DPI has the Framed IP address of the UE; this is all which is needed to behave as an SBC to modify the QoS.

The two choices are technically possible and the Rxbased implementation is more logical.

10.1.5 Traces for the charging by an external DPI

10.1.5.1 Data provided by the GGSN-PGW which may be used for charging

These data are recovered from the Create PDP Context (3G) or the Create Session (4G);

ASN1 Profil decoded:

```
---------- PDU numero 1 -------------
value InsertSubscriberDataArg ::=
{
   subscriberStatus serviceGranted,
   teleserviceList
   {
     short Message MO-PP,
     short Message MT-PP
   },
   gprsSubscriptionData
   {
     completeDataListIncluded NULL,
     gprsDataList
     {
         {
           pdp-ContextId 3,
           pdp-Type IETF allocated address: IPv4,
```

```
                qos-Subscribed '094205'H,
                apn mms.malitel4.ml,
                ext-QoS-Subscribed '023101EAEA1105EAEA'H,
                pdp-ChargingCharacteristics '0400'H
             }
          }
       },
       networkAccessMode bothMSCAndSGSN,
       chargingCharacteristics '0400'H
    }
```

```
---------- PDU numero 2 -------------
value InsertSubscriberDataArg ::=
{
   gprsSubscriptionData
   {
      gprsDataList
      {

         {
           pdp-ContextId 2,
           pdp-Type IETF allocated address: IPv4,
           qos-Subscribed '099112'H,
           apn wap.malitel3.ml,
           ext-QoS-Subscribed '017296EAEA7505EAEA'H,
           pdp-ChargingCharacteristics '0400'H
         },

         {
           pdp-ContextId 1,
           pdp-Type IETF allocated address: IPv4,
           qos-Subscribed '099112'H,
           apn web.malitel3.ml,
           ext-QoS-Subscribed '017296EAEA7505EAEA'H,
           pdp-ChargingCharacteristics '0400'H
         }
      }
   }
}
```

1) in the 1rst PDU

```
chargingCharacteristics '0400'H // Global for all APNs
```

2) and in each APN including the one used in the session created

```
apn mms.malitel4.ml
pdp-ChargingCharacteristics '0400'H // TS 32.298 and TS 32.251
```

The *Charging Characteristics* paramter is 2 octets ***operator specific(!)*** and does not contain any applicable information on the charging plan of the subscriber only flags for Home default, Visiting default, Roaming default, specific APN, and Specific Subscription (without indication about the applied plan).

The Charging Characteristics paramter comes from the HLR's profile for that particular subscriber

$\rightarrow$ to be set by the HLR-HSS but not usable for that purpose.

In addition:

```
         Framed-IP-Address (Source or Destination IP allocated to the
         UE's traffic)
         IMSI
         MSISDN
         RAT type (UTRAN, GERAN, eUTRAN)
         IMEI (not normally used)
         NAS-IP-Address (IP of the GGSN-PGW, not normally used)
         SGSN or MME IP (can be used for charging)
```

In 3G and 4G:

```
         User Location Info (the SAI visited by the subscriber
containing the PLMN ID at the begining)-> location dependent charging)
208.1.790.62319
```

In 4G only:

```
Serving Network (4G only MCC-MNC visited)
```

Conclusion of this analysis for data charging

If specific rates are applied for the same visited network for different subscribers and the same destination site, it must be provisioned in the OCS: nothing in the data available to the GGSN-PGW allows providing an information of the rating plan of a particular subscriber.

10.1.5.2 RADIUS interface between GGSN-PGW and PCEF-DPI (which is the radius server)

The GGSN-PGW sends the RADIUS Accounting-Request Start to the external PCEF-DPI which *creates contexts that it will access by the IP of the UE* when it receives a new *Framed-IP address*. This *context is freed* by the external PCEF-DPI when a RADIUS Accounting-Request (Stop) is received or when it generates a RADIUS Disconnect with the same Radius *Acct-Session-Id*.

The PCEF-DPI analyzes the Gi/SGi IP flow to recognize new URLs (IP) *with a destination IP table* and depending on the destination address sends

CCR/Gy/Diameter requests to the OCS (TS 32.299) will have to modify the relevant AVPs (partial list) Gy:

1. Rating-group AVP
2. Multiple-services-credit-control

can be used to change the rating based on a destination IP table in the external PCEF-DPI.

10.1.5.3 Content of the RADIUS accounting-request (start) received by the PCEF-DPI

```
- - - - Super Detailed GTP, SIP, DIAMETER, RADIUS-> Analyser (C)HALYS
      - - - -
- - -message length = 207
   ____Header RADIUS-> (port 5062): SENT ____
  RADIUS-> Command: (4):Accounting-Request
  Packet identifier 4C
  headerRADIUS(with Authenticator)+payload = 207
  Authenticator 861206747B4BDBAC5A679F9DD2DBD014 (MD5 calculated using
      shared
secret: totoalaplage)

HEXA payload
0102060600000002280600000012C0B3533363837363333372D06000000010806C7FF
070A0706000000071E1174656C6E616D6F62696C652E636F6D1F0D3333373537373931
3036351A17000028AF0111323038393430303030303030313036351A0C000028AF0606C0
A802021A09000028AF0A03351A18000028AF141233353233313530363138333439320
311A10000028AF160A0102F8100316F36F1A0C000028AF1506000000001050600000000
2906000000000406C0A80272
(1):User_Name(e.g.User Name=Josef set in the softphone, not
necessarily the telephone number! not used for login OpenSIP)
 length AVP = 2
L.0440]: value =
 (6):Service-Type
 length AVP = 6
 (2):Framed-User
 (40):Acct-Status-Type(Start,Stop,Interim-Update,etc..)
 length AVP = 6
 (1):Start
 (44):Acct-Session-Id(Call ID of the session) // to be saved in the
PCEF-DPI context to correlate with the Stop or Disconnect
 length AVP  =  11
 value = 536876337 // = Session ID (PCRF)
 (45):Acct-Authentic
 length AVP = 6
 (1):RADIUS
 (8):Framed-IP-Address(IP of handset assigned by DHCP);presence means
 Terminal is IP ACTIVATED!
 length AVP = 6
 value = 199.255.7.10 //Framed IP AddressASSIGNED TO UE
 (7):Framed-Protocol
 length AVP = 6
```

```
 (7):GPRS-PDP-Context
 (30):Called-station-Id(APN selected by handset(GPRS) or MAC adress of
Hotspot(WiFi)
 length AVP = 17
 value = telnamobile.com // APN of UE for charging
 (31):Calling-Station-Id:MSISDN of handset(GPRS or EAP-SIM) or MAC
address of Terminal(Hotspot WiFi)
 length AVP = 13
 MSISDN = 33757791065 // MSISDN of UE for charging
 (26):Vendor-Specific(3GPP TS 29.061)
 Vendor_ID = 10415((1):3GPP-IMSI of handset)
 IMSI = 208940000001065 // IMSI of UE for charging
 (26):Vendor-Specific(3GPP TS 29.061)
 Vendor_ID = 10415((6):3GPP-SGSN-IP Address(visited by handset))
 value = 192.168.2.2
 (26):Vendor-Specific(3GPP TS 29.061)
 Vendor_ID = 10415((10):3GPP-NSAPI)
 value = 5
 (26):Vendor-Specific(3GPP TS 29.061)
 Vendor_ID = 10415((20):3GPP-IMEISV(IMEI+SW Version))
 IMEISV = 3523150618349201 // IMEI for optional charging criteria
 (26):Vendor-Specific(3GPP TS 29.061)
 Vendor_ID = 10415((22):User Location Info)
 type of ULI: (1):Service Area Identity(SAI) MCC.MNC.LAC.SAC(2 octets)
 ULI = 208.1.790.62319 // ULI for optional charging criteria
 (26):Vendor-Specific(3GPP TS 29.061)
 Vendor_ID = 10415((21):3GPP-RAT(Radio Access Technology Type))
 RAT = (1):UTRAN
 (5):NAS-Port(physical connection on NAS,NOT the TCP or UDP 'port
number')
 length AVP = 6
 value = [unhandled integer length(0)]
 (41):Acct-Delay-Time(delay added in the NAS for Accounting-Request
NAS-> AAA) in seconds
 length AVP = 6
 value = 0
 (4):NAS-IP-Address(source of the RADIUS Request: GGSN , Hotspot WiFi,
OpenSIP )
 length AVP = 6
 value = 192.168.2.114 // IP of the GGSN-PGW
```

10.1.5.4 Content of the RADIUS accounting-request (stop) receive by the PCEF-DPI

Note: No Interim-Update generated

```
_____Header RADIUS->
(port 5062): SENT _____
RADIUS-> Command: (4):Accounting-Request
Packet identifier DD
headerRADIUS(with Authenticator)+payload = 237
Authenticator 3997D88EA0F0034C178651CD9C4E93AB (MD5 calculated using
    shared
```

```
secret: totoalaplage)
HEXA payload
0102060600000000022806000000022A060001D4D62B06000158762F060000015D30060
00001772E06000008292C0B3533363837363333372D06000000010806C7FF070B0706
000000071E1174656C6E616D6F62696C652E636F6D1F0D333373537373931303635
A17000028AF011132303839343030303030303031303635A1A0C000028AF0606C0A80202
1A09000028AF0A03351A18000028AF1412333532333135303631383334393230311A1
0000028AF160A0102F810031619CD1A0C000028AF15060000000010506000000002906
000000000406C0A80272
```

```
 (1):User_Name(e.g.User Name=Josef set in the softphone, not
necessarily the telephone number! not used for login OpenSIP)
 length AVP = 2
 value =
 (6):Service-Type
 length AVP = 6
 (2):Framed-User
 (40):Acct-Status-Type(Start,Stop,Interim-Update,etc..)
 length AVP = 6
 (2):Stop // Free the IP context in the PCEF-DPI
 // counts of the entire session
 (42):Acct-Input-Octets(NAS-> Terminal CUMULATED)        DL
 length AVP = 6
 value = 120022
 (43):Acct-Output-Octets(Terminal-> NAS CUMULATED)       UL
 length AVP = 6
 value = 88182
 (47):Acct-Input-Packets(NAS-> Terminal CUMULATED)       DL
 length AVP = 6
 value = 349
 (48):Acct-Output-Packets(Terminal-> NAS CUMULATED)      UL
 length AVP = 6
 value = 375
 (46):Acct-Session-Time(Terminal<-> NAS CUMULATED seconds)
 length AVP = 6
 value = 2089
 (44):Acct-Session-Id(Call ID of the session)
 length AVP = 11
 value = 536876337 // = Session ID (PCRF)
```

10.2 Standard 3GPP user data provisioning: LDAP and SOAP

10.2.1 3GPP user data repository architecture

There is an LDAP server with a central database. The provisioning gateway and all the equipment which need it are LDAP clients of this UDR. The individual databases in the HLR-HSS and the PCRF are caches of this UDR.

The HLR-HSS uses the Ud protocol [10.1] which includes SOAP notifications received from the UDR and LDAP SEARCH to obtain the

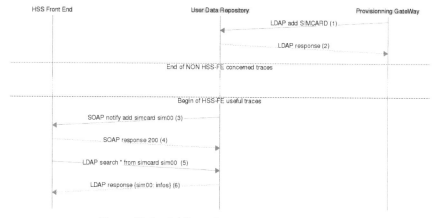

Figure 10.6 Adding a SIM card in the HLR-HSS.

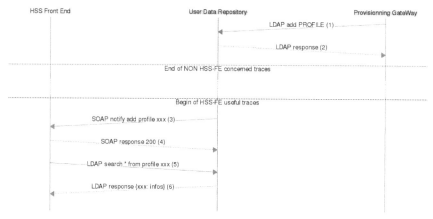

Figure 10.7 Profile creation.

corresponding information and stores it in the HLR-HSS local cache database.

This is not a very accurately defined function the role of which is the same as that of the UDM in 5G [12.7].

The UDR can be the main database of the HLR-HSS or simply an intermediate partial database.

Figures 10.6 and 10.7 show the addition of an IMSI and of a profile in the UDR. The HLR-HSS then receives a SOAP Notify from the UDR and reads the modified or added data.

Figure 10.8 is the addition of a subscriber with a reference to its IMSI and profile. In practice, the UDR architecture is not used and there are

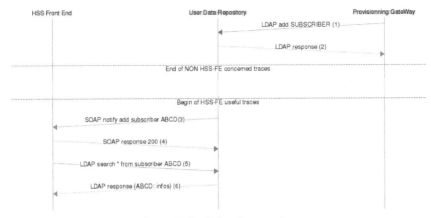

Figure 10.8 Subscriber creation.

proprietary interfaces between the HLR-HSS and the CRM which is the "provisioning gateway." There is also the PCRF to provision as it needs the QoS subscription data as many MNOs control the QoS only with the PCRF, while the QoS data in the HLR-HSS are meaningless.

References

[10.1] 3GPP TS 29.335 V14.0.0 (2017-04), "Digital cellular telecommunications system (Phase 2+) (GSM); Universal Mobile Telecommunications System (UMTS); LTE; User Data Convergence (UDC); User data repository access protocol over the Ud interface; Stage 3," Release 14, *the chapter explains this standard architecture.*

[10.2] 3GPP TS 23.008 V14.4.0 (2018-01), "Digital cellular telecommunications system (Phase 2+) (GSM); Universal Mobile Telecommunications System (UMTS); LTE; Organization of subscriber data," Release 14.

[10.3] 3GPP TS 29.212 v12.10.0 (2015-10), "Universal Mobile Telecommunications System (UMTS); LTE; Policy and Charging Control (PCC); Reference points," Release 12. *Protocol Gx between GGSN-PGW (lPCEF) and the PCRF.*

[10.4] 3GPP TS 129 214 V12.9.0 (2015-10), "Universal Mobile Telecommunications System (UMTS); LTE; Policy and charging control over Rx reference point," Release 12, *Interface Rx between PCRF and ASs.*

[10.5] O. Iteanu, "CLOUD ACT : le Congrès américain a-t-il voté contre le RGPD?", SD Magazine, June 6, 2018.

11

Appendix: Detailed Traces for the Different Chapters

11.1 Triggering the IP connection of the SIM IP to the OTA-IP server (Chapter 4)

11.1.1 OTA-IP configuration of the SIM for TCP (HTTPs) for UDP (BIP/CAT-TP)

The configuration file in Figure 11.1 must be present for OTA-IP; the identifier and detailed content is somewhat proprietary for a given SIM manufacturer, and has been presenting in most SIM cards since 2010. It may be provided directly by the manufacturer at order or later updated by OTA SMS. This file is mandatory for HTTPs or BIP/CAT-TP whether it is network initiated by an SMS-MT or SIM initiated with a poll Applet. For SIM card initiated connection, *the IP of the OTA server must be initialized in the parameter Data Destination Address, which allows the Port;* otherwise, the parameters can also be overwritten and provided in the triggering SMS-MT of the next section which uses 192.168.2.118.

When the OTA-IP configuration file is ready, either:

- the OTA server can send an SMS-MT with a payload described in the following sections,
- or the "Applet for OTA pull" of Figure 4.1 can send a command to the SIM card IP client.

11.1.2 Triggering the OTA-IP connection from the "Applet for OTA pull" or with an SMS-MT

The connection to the OTA-IP server must include an identifier of the customer in the first PDU so that the OTA server will know if OTA commands

File Identifier 3F00/6F55 (Manufacturer J)	File structure type: Transparent		Mandatory fo BIP/CAT-TP	
File size: 100 bytes			Update activity: low	
Acces conditions READ ADM UPDATE ADM				
Tag Value	Description	Real example	M/C/O	Length
0x99	TAG OPEN CHANNEL PARAMETERS	99	M	1 byte
	Length of file	33 hex (51 decimal)	M	
0x05	Alpha ID (TLV)	05 00	C	
0x35	Bearer Description (TLV)	35 07 02 00 00 03 00 00 02	M	7 bytes
0x39	Buffer Size (TLV)	39 02 05 78	M	2 bytes
0x0D	Login (TLV)	0D 06 04 48 61 6C 79 73 DCS H a l y s	C	
0x0D	Password (TLV)	0D 06 04 48 61 6C 79 73 DCS H a l y s	C	
0x47	Network Address Name(APN) (TLV)	47 06 04 48 61 6C 79 73 DCS H a l y s	C	
0x3C	Interface Transport Level (server port) (TLV)	3C 03 02 **1F 52** (port 8018 in decimal) (02=TCP HTTPS)	C	
0x3E	Data Destination Address (TLV) sent in the SYN initial	3E 05 21 C0 A8 02 76 (IP 192.168.2.118 in decimal)	C	

Figure 11.1   Example of content of the proprietary configuration 3F00/6F55 file (Manufacturer J) in the SIM to enable TCP(HTTPS) or UDP(BIP/CAT-TP).

must be sent for that customer. Usually the ICCID is used as it is known both from the "Applet for OTA pull" (SIM card initiated OTA-IP) and from the OTA-IP server if SMS-MT push is used. Several subscribers would be in an OTA process and an identifier for the first connection setup IP message is mandatory. Then the other messages are related by the IP Source address of the received packets.

The solution is that the internal command "Applet for OTA pull" → IP client in the SIM or the SMS-MT push includes the ICCID 8933940000000000171 of the IMSI 208940000001016 used in the example below. The OTA server data base of Figure 11.2 has the correspondance ICCID-IMSI.

Figure 11.2 Details of the OTA server database used for ICCID → MSISDN, IMSI, and IMEI.

11.1.3 Details of the SMS-MT to network initiate a forced BIP/CAT-TP (IP UDP connection to the OTA server while the SIM configuration is IP TCP (HTTPs)

11.1.3.1 Coding of the SMS-MT payload (ETSI TS 102 223 [4.11]) to open the BIP and CAT-TP channels

This is a tutorial and the purpose of this command is to read the active IMSI while forcing with an SMS-MT trigger the use of OTA-IP UDP while HTTPS (TCP) was default configured in the SIM cards.

The use of the payload sent in an SMS-MT by the OTA server is detailed in Section 4.5.3.

The payload is described in ETSI TS 102 223 [4.11] and the security of the trigger sending by SMS in TS 23.048 [4.5]. The Tags such as "35" bearer description tag are all in ETSI TS 102 220 [4.17].

Below is the payload to open a BIP/CAT-TP session between the SIM card and the OTA server. That payload is sent by the OTA server to the card by SMS and can override any parameter in Figure 13.1 where TCP HTTPS = 02 is replaced by BIP/CAT-TP = **01** because the SIM card has only BIP/CAT-TP (HTTPS requires more expensive SIMs in general).

80EC0101**19**350702000003000002390204003E0521C0A802763C03**011F52**
80EC0102**1**13C03000004360A 8933940000000000171

The green part are the commands to set the BIP channel SIM ↔ UE and the blue part to send end-to-end CAT-TP (type port=**01**) SIM ↔ OTA server.

BIP channel setting command part

80: CLA

EC: INS

01: P1

01: P2 (= Request for BIP channel opening)

19: length of the command (25)

35: BEARER DESCRIPTION tag

07: BEARER DESCRIPTION length

02: bearer type (= reserved for GSM/3GPP)

000003000002: bearer parameters (Precedence Class (Priority), Delay Class, Reliability Class, Peak Throughput, Mean Throughput, PDP IP)

39: BUFFER SIZE tag

02: BUFFER SIZE length

0400: 1024 bytes

3E: DESTINATION ADDRESS tag

05: DESTINATION ADDRESS length

21: type of address (= IPv4)

C0A80276 This is the IP 192.168.2.118 *of the OTA server* to which the SIM connects to using the CAT-TP protocol.

3C: PORT CARD → OTA SERVER tag

03: PORT length

02: type port (02 = TCP (HTTPS OTA IP), 01 = UDP (BIP/CAT-TP) overrides TCP configured in the card

1F 52 (port number (UDP 8018 decimal chosen arbitrarily by the OTA server which it listens to)

CAT-TP channel setting command part

80: CLA

EC: INS

01: P1

02: P2 (= Request for CAT_TP link establishment)

11: length of the command

3C: UICC/terminal interface transport level tag

03: UICC/terminal interface transport level length

00: Transport protocol type (= reserved)

0004: Port number

36: CHANNEL DATA tag

0A: CHANNEL DATA length

8933940000000000171: CHANNEL DATA string (returned in the SYN of the connection SIM-> server)

The payload is used in the MAPPN_sm_rp_ui parameter of the SMS-MT.

11.1.3.2 SMS-MT parameters details (TS 23.040 [4.20])

```
--------------------- HDR ------------------------ [ Fri Feb 2
    15:22:27.64 2018
- - - - Super Detailed SS7 Analyser
      PA_Len = 150
      MAP-MT-FORWARD-SHORT-MESSAGE-REQ(Version 2+)(69)
          MAPPN_timeout(45)
            L = 002
            Data: timeout value = 51 sec
          MAPPN_invoke_id(14)
            L = 001
            Data: 1
          MAPPN_sm_rp_da(23)
            L = 019
            Data: TA_SC_ADR
                  Ext = No extension
                  Ton = International
                  Npi = ISDN
                  Address = 208940000001016
          MAPPN_sm_rp_oa(24)
            L = 009
            Data: TA_SC_ADR
                  Ext = No extension
                  Ton = International
                  Npi = ISDN
                  Address = 33970670001
          MAPPN_sm_rp_ui(25)
            L = 107
            Data: Message Type = SMS_DELIVER(SMS-MT)
                  TP_RP = Request for reply path
                  TP_UDHI= Header in TP-User-Data
                  TP_SRI = A status report will not be returned to the
                    SME
                  TP_VPF = TP_VP field not present
                  TP_MMS = No more messages are waiting for the MS in
                    this SC
                  Originating mobile address =
                     Type of number = International
                     Numbering Plan = ISDN
                     Address = 33970670001
                  TP-Protocol-Identifier = 7F (SIM Data upload or
                    download)
                  TP-Data-Coding-Scheme = F6
                     Message Class = Class 2(SIM Card)
                     Alphabet = 8 bit(image/sound/download)//
```

```
                              SMS sent to the SIM

                  TP_Service_Centre_Time_Stamp = 18.02.02 18:25:40 04
                  TP-User-Data-length = 79
                     Length TP_User Data Header = 2
                         IEI = 112 (SIM Command Packet Identifier) (70
                         hexa)
                            IEI Data Length = 0
                  Command Packet Length=69 Command Header Length=21
                  Security Parameter Indicator:
                   SPI1(12):
                    Cryptographic Checksum(CC)
                    No Ciphering
                    Process if counter>RE
                   SPI2(21):
                    PoR to be sent to Sending Entity
                    No Security applied to PoR
                    PoR not ciphered
                    PoR by SUBMIT
                   KIc(15)
                    DES
                    Triple DES 2 different keys
                    Keyset number(KIc-KID-KIK) used 1
                   KID(15)
                    DES
                    Triple DES 2 different keys
                    Keyset number(KIc-KID-KIK) used 1
              <?xml version="1.0" encoding="ISO-8859-1"?>
              <SIM_parameters>
                  <TAR>B00010</TAR>
                  <RemoteFileManager>Manufacturer J</RemoteFileManager>
                  <CNTR>0000000067</CNTR>
                  <NumberOfPaddingOctetsAtTheEnd>0</
         NumberOfPaddingOctetsAtTheEnd>
                  <Crypto>CC or DS = 142F008A1F666546</Crypto>
                  TP-User-Data =

 "00451512211515B000100000000006700142F008A1F66654680EC010119 35070200000
 3000002390204003E0521C0A802763C03011F5280EC0102113C03000004360A8933940
 000000000171"
```

One sees the payload coded with ETSI TS 102.223 [4.11] described in Section 4.5.3 which opens the BIP channel and the CAT-TP channel and sets in orange the TS 23.048 [4.5] security parameters.

11.1.3.3 Opening of the IP channel by the SIM card (extracted from Figure 4.7)

```
    <--- IP UDP SYN pdu opening 15:22:31.29
         payload identification = Channel Data = 8933940000000000171
 = ICCID
```

```
----- IP UDP SYN + ACK pdu
    -------------------------------------------------→
15:22:31.29
```

With the ICCID, the OTA server is able to find the MSISDN +33757791016 of the subscriber data to create an SMS-MT and find the pending commands to be sent to the SIM.

11.1.3.4 Confirmation of the reception of triggering SMS-MT by the SIM and of the establishment of the IP UDP channel to the OTA IP server

```
---------------------- HDR ------------------------- [ Fri Feb 2
15:22:31.51 2018
- - - - Super Detailed SS7 Analyser
        PA_Len = 12
        MAP-MT-FORWARD-SHORT-MESSAGE-CNF(Version 2+)(192)
          MAPPN_invoke_id(14)
            L = 001
            Data: 1
          MAPPN_sm_rp_ui(25)
            L = 005
            Data: Message Type = SMS_DELIVER_REPORT
                (Hex) 00077FF600
                    TP-UDHI = No Header in TP-User-Data
                    TP-MMS = More messages are waiting for the MS in
                     this SC

                    TP-PI = 07 TP-UDL, TP-DCS and TP-PID all present
                    TP-Protocol-Identifier = 7F (SIM Data upload or
                     download)
                    TP-Data-Coding-Scheme = F6
                       Message Class = Class 2(SIM Card)
                       Alphabet      = 8 bit(image/sound/download)
                    TP-User-Data-length = (0) no User-Data
```

At this stage, the SIM card will attempt in this IP UDP case to establish a BIP channel with the UE and *then a CAT-TP IP connection with the OTA IP server.*

11.1.3.5 OTA-IP sequence of commands (read IMSI)

```
<------- IP UDP ACK pdu ----------------------------- 15:22:31.71

-------- -IP UDP ACK pdu (TS 51.011) SELECT EFimsi, IMSI ------>
15:22:31.79

payload 01301512011515 B0001000000003E8005322123717290B9C A0A40000023F
```

```
00A0A40000027F20 A0A40000026F07A0B0000009 (security TS 102 225)
ciphering not mandatory (not used here)
```

11.1.3.6 SMS-MO received by the OTA server with the PoR

When the OTA is completed (the SIM has been updated or read) of failed, it sends a PoR (an SMS-MO) to the OTA server which had initiated the operation to give the result.

```
---------------------- HDR ------------------------ [ Fri Feb 2
    15:22:33.81 2018
    - - - - Super Detailed SS7 Analyser
    PA_Len = 83
    MAP-FORWARD-SHORT-MESSAGE-IND(Version 2)(4)
      MAPPN_invoke_id(14)
        L = 001
        Data: 1
      MAPPN_sm_rp_da(23)
        L = 009
        Data: TA_SC_ADR
              Ext = No extension
              Ton = International
              Npi = ISDN
              Address = 33970670000
      MAPPN_sm_rp_oa(24)
        L = 017
        Data: TA_SC_ADR
              Ext = No extension
              Ton = International
              Npi = ISDN
              Address = 33757791016 c4149503f947a54d
      MAPPN_sm_rp_ui(25)
        L = 036
        Data: Message Type = SMS_SUBMIT(SMS-MO)
              TP_RP = No request for reply path
              TP_UDHI= Header in TP-User-Data
              TP_SRR = Status Report not requested
              TP_VPF = TP_VP field not present
              TP_RD = Accept SMS with same TP_MR
              TP_Message reference = 1
              Destination mobile address =
                  Type of number = International
                  Numbering Plan = ISDN
                  Address = 33970670001
              TP-Protocol-Identifier = 00 (No Interworking: SME-to-SME
                protocol)
              TP-Data-Coding-Scheme = F6
                  Message Class = Class 2(SIM Card)
                  Alphabet      = 8 bit(image/sound/download)
              TP-User-Data-length = 23
                  Length TP_User Data Header = 2
```

```
                    IEI = 113 (SIM Response Packet Identifier)(71 hexa)
                       IEI Data Length = 0
                 Command Packet Length=18 Command Header Length=10
           <?xml version="1.0" encoding="ISO-8859-1"?>
           <SIM_parameters>
               <TAR>B00010</TAR>
               <RemoteFileManager>Jmanufacturer</RemoteFileManager>
               <CNTR>0000000067</CNTR>
               <NumberOfPaddingOctetsAtTheEnd>0</NumberOfPadding
                OctetsAtTheEnd>
               <ResponseStatusCode>(00): PoR OK</ResponseStatusCode>
               <Crypto>No RC,CC or DS</Crypto>
               <NumberOfCommandsExecuted>2</NumberOfCommandsExecuted>
               <StatusConditionsReturnedBySIM>
                   <SW1>90</SW1>
                   <SW2>00</SW2>
               </StatusConditionsReurnedbySIM>
               <Info>SUCCESSFUL: normal ending of the command</Info>
               <LastCommand>80B00000</LastCommand>
               <READ BINARY>2
           </SIM_parameters>

       MAPPN_imsi(18)
         L = 008
         Data: Address = 208940000001016
- - - - - - - - - - - - - - - - - - - -- - - - - - - - - - - - - -
```

```
---------------------- HDR ------------------------ [ Fri Feb 2
15:22:33.94 2018
- - - - Super Detailed SS7 Analyser
     PA_Len = 83
     MAP-FORWARD-SHORT-MESSAGE-IND(Version 2)(4)
       MAPPN_invoke_id(14)
         L = 001
         Data: 1
```

If the OTA- IP connection had failed, the PoR would indicate a code "technical problem."

11.1.3.7 Completion and closing of the OTA-IP session

```
<------- IP UDP ACK pdu -------------------------- 15:22:34.70

<------- IP UDP ACK pdu (PoR TS 51.011) IMSI and OK------
    15:22:35.50

payload 01170AB0001000000003E8000004 9000 082906601096928991 (IMSI
206060169290819)

-------- IP UDP ACK --------------------------> 15:22:35.50
```

```
                  other READ or UPDATE commands

----------➞IP UDP RST pdu Close connexion---------------------------->

              or Time out(15 sec about), no more commands
received by the SIM

<---------- IP UDP RST Close Channel ----------------------------
    15:22:51.26
```

Figure 11.3 gives the high level PCAP traces of the sending of the 4G SMS-MT which triggers the establishment of the BIP and CAT-TP channels by the SIM card.

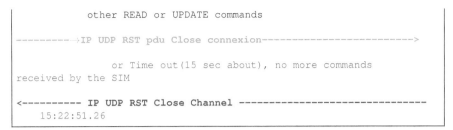

```
15 cmd=UnknownRequest(8388646) flags=R--- appl=Unknown(16777313) h2h=1e361f e2e=42c0a56c
16 id-Paging
17 id-initialUEMessage, Service request
18 SACK id-InitialContextSetup, InitialContextSetupRequest
19 SACK id-UECapabilityInfoIndicationUECapabilityInformation
20 id-InitialContextSetup, InitialContextSetupResponse
21 Modify Bearer Request
22 Modify Bearer Response
23 id-downlinkNASTransport, Downlink NAS transport(DTAP) (SMS) CP-DATA (RP) RP-DATA (Network to MS)
24 SACK id-uplinkNASTransport, Uplink NAS transport(DTAP) (SMS) CP-ACK
25 id-uplinkNASTransport, Uplink NAS transport(DTAP) (SMS) CP-DATA (RP) RP-ACK (MS to Network)
26 id-downlinkNASTransport, Downlink NAS transport(DTAP) (SMS) CP-ACK
27 cmd=UnknownAnswer(8388646) flags=-P-- appl=Unknown(16777313) h2h=1e361f e2e=42c0a56c
28 id-UEContextReleaseRequest, UEContextReleaseRequest
29 Release Access Bearers Request
30 Release Access Bearers Response
31 SACK id-UEContextRelease, UEContextReleaseCommand
32 SACK id-UEContextRelease, UEContextReleaseComplete
```

Figure 11.3   FW-SM-MT call flow S1-AP eNodeB-MME, SGd MME-SMSC and GTP-C MME-S/PGW.

We have effectively been able to read the active IMSI 206060169290819 and OTA-IP UDP was used while OTA-IP TCP (HTTPs) was configured in the SIM. This IMSI is an auxilary IMSI because the nominal one could not find an agreeing network coverage.

11.2 SMS-MT with diameter SMSC → UE (Chapter 5)

Trace details

```
No.

    15 cmd=UnknownRequest(8388646) flags=R--- appl=MT_FORWARD_
SHORT-MSG(16777313) h2h=1e361f e2e=42c0a56c

Frame 15: 408 bytes on wire (3264 bits), 408 bytes captured
(3264 bits)
Linux cooked capture
```

```
Internet Protocol Version 4, Src: 192.168.2.114 (192.168.2.114), Dst:
192.168.2.113 (192.168.2.113)
Stream Control Transmission Protocol, Src Port: 3871 (3871), Dst Port:
    3871
(3871)
Diameter Protocol
    Version: 0x01
    Length: 344
    Flags: 0x80
    Command Code: 8388646 MT-FORWARD_SHORT_MSG Request
    ApplicationId: SGs (16777313)
    Hop-by-Hop Identifier: 0x001e361f
    End-to-End Identifier: 0x42c0a56c
    AVP: Session-Id(263) l=34 f=-M- val=smscHalys;1493279010;35571
    AVP: Auth-Session-State(277) l=12 f=-M- val=NO_STATE_
    MAINTAINED (1)
    AVP: Origin-Host(264) l=46 f=-M- val=smsc.epc.mnc094.mcc208.
    3gppnetwork.org
    AVP: Origin-Realm(296) l=41 f=-M- val=epc.mnc094.mcc208.
    3gppnetwork.org
    AVP: Destination-Host(293) l=45 f=-M- val=mme.epc.mnc094.mcc208.
    3gppnetwork.org
    AVP: Destination-Realm(283) l=41 f=-M- val=epc.mnc094.mcc208.
    3gppnetwork.org
    AVP: User-Name(1) l=23 f=-M- val=260060169299766
    // SM-RP-DA = IMSI
    AVP: Unknown(3300) l=23 f=VM- vnd=TGPP val=3333393730363730303037
    // SC-Addess GT
    AVP: Unknown(3301) l=43 f=VM- vnd=TGPP val=040b913379605701f200007
    ...
    // SM-RP-UI

No.
    16 id-Paging

Frame 16: 108 bytes on wire (864 bits), 108 bytes captured (864 bits)
Linux cooked capture
Internet Protocol Version 4, Src: 192.168.2.113 (192.168.2.113),
Dst: 192.168.2.117 (192.168.2.117)
Stream Control Transmission Protocol, Src Port: s1-control (36412),
Dst Port: s1-control (36412)
S1 Application Protocol

No.
    17 id-initialUEMessage, Service request

Frame 17: 124 bytes on wire (992 bits), 124 bytes captured (992 bits)
Linux cooked capture
Internet Protocol Version 4, Src: 192.168.2.117 (192.168.2.117),
Dst: 192.168.2.113 (192.168.2.113)
Stream Control Transmission Protocol, Src Port: s1-control (36412),
Dst Port: s1-control (36412)
S1 Application Protocol
```

```
No.
    18 SACK id-InitialContextSetup, InitialContextSetupRequest

Frame 18: 192 bytes on wire (1536 bits), 192 bytes captured
(1536 bits)
Linux cooked capture
Internet Protocol Version 4, Src: 192.168.2.113 (192.168.2.113),
Dst: 192.168.2.117 (192.168.2.117)
Stream Control Transmission Protocol, Src Port: s1-control (36412),
Dst Port: s1-control (36412)
S1 Application Protocol

No.
    19 SACK id-UECapabilityInfoIndicationUECapabilityInformation

Frame 19: 296 bytes on wire (2368 bits), 296 bytes captured
(2368 bits)
Linux cooked capture
Internet Protocol Version 4, Src: 192.168.2.117 (192.168.2.117),
Dst: 192.168.2.113 (192.168.2.113)
Stream Control Transmission Protocol, Src Port: s1-control (36412),
Dst Port: s1-control (36412)
S1 Application Protocol

No.
    20 id-InitialContextSetup, InitialContextSetupResponse

Frame 20: 108 bytes on wire (864 bits), 108 bytes captured (864 bits)
Linux cooked capture
Internet Protocol Version 4, Src: 192.168.2.117 (192.168.2.117),
Dst: 192.168.2.113 (192.168.2.113)
Stream Control Transmission Protocol, Src Port: s1-control (36412),
Dst Port: s1-control (36412)
S1 Application Protocol
```

The Modify Bearer Request below is really not useful as the procedure uses only the control plane.

```
No.
    21 Modify Bearer Request
Frame 21: 101 bytes on wire (808 bits), 101 bytes captured (808 bits)
Linux cooked capture
Internet Protocol Version 4, Src: 192.168.2.113 (192.168.2.113),
Dst: 192.168.2.114 (192.168.2.114)
User Datagram Protocol, Src Port: gtp-control (2123),
Dst Port: gtp-control (2123)
GPRS Tunneling Protocol V2
Modify Bearer Request

No.
    22 Modify Bearer Response
```

```
Frame 22: 62 bytes on wire (496 bits), 62 bytes captured (496 bits)
Linux cooked capture
Internet Protocol Version 4, Src: 192.168.2.114 (192.168.2.114),
Dst: 192.168.2.113 (192.168.2.113)
User Datagram Protocol, Src Port: gtp-control (2123),
Dst Port: gtp-control (2123)
GPRS Tunneling Protocol V2
Modify Bearer Response

No.
     23 id-downlinkNASTransport, Downlink NAS transport(DTAP) (SMS)
CP-DATA (RP) RP-DATA (Network to MS)

Frame 23: 148 bytes on wire (1184 bits), 148 bytes captured
(1184 bits)
Linux cooked capture
Internet Protocol Version 4, Src: 192.168.2.113 (192.168.2.113),
Dst: 192.168.2.117 (192.168.2.117)
Stream Control Transmission Protocol, Src Port: s1-control (36412),
Dst Port: s1-control (36412)
S1 Application Protocol
    S1AP-PDU: initiatingMessage (0)
        initiatingMessage
            procedureCode: id-downlinkNASTransport (11)
            criticality: ignore (1)
            value
                DownlinkNASTransport
                    protocolIEs: 3 items
                        Item 0: id-MME-UE-S1AP-ID
                        Item 1: id-eNB-UE-S1AP-ID
                        Item 2: id-NAS-PDU
                            ProtocolIE-Field
                                id: id-NAS-PDU (26)
                                criticality: reject (0)
                                value
                                NAS-PDU:
27e4cd5f2b0507622e29012b010207913379600700f7001f...

                                    Non-Access-Stratum (NAS)PDU
                                        0010 .... = Security header type:
Integrity protected and ciphered (2)
                                            .... 0111 = Protocol
discriminator: EPS mobility management messages (0x07)
                                        Message authentication code:
0xe4cd5f2b
                                        Sequence number: 5
                                        0000 .... = Security header type:
Plain NAS message, not security protected (0)
                                            .... 0111 = Protocol
discriminator: EPS mobility management messages (0x07)
                                        NAS EPS Mobility Management
Message Type: Downlink NAS transport (0x62)
```

```
                                              NAS message container
                                                 Length: 46
                                                 NAS message container content:
29012b010207913379600700f7001f040b913379605701f2...
                                              GSM A-I/F DTAP - CP-DATA
                                              GSM A-I/F RP - RP-DATA
                                              (Network to MS)
                                              GSM SMS TPDU (GSM 03.40)
                                              SMS-DELIVER
                                                 0... .... = TP-RP:
TP Reply Path parameter is not set in this SMS SUBMIT/DELIVER
                                                 .0.. .... = TP-UDHI:
The TP UD field contains only the short message
                                                 ..0. .... = TP-SRI:
A status report shall not be returned to the SME
                                                 .... .1.. = TP-MMS:
No more messages are waiting for the MS in this SC
                                                 .... ..00 = TP-MTI:
SMS-DELIVER (0)
                                                    TP-Originating-
                                                    Address -
                                                    (33970675102)
                                                    TP-PID: 0
                                                    TP-DCS: 0
                                                    TP-Service-Centre-
                                                    Time-Stamp
                                                    TP-User-Data-Length:
(13) depends on Data-Coding-Scheme
                                                    TP-User-Data
                                                       SMS text:
```

Test inactif6

```
No.
    24 SACK id-uplinkNASTransport, Uplink NAS transport(DTAP)
(SMS) CP-ACK

Frame 24: 140 bytes on wire (1120 bits), 140 bytes captured
(1120 bits)
Linux cooked capture
Internet Protocol Version 4, Src: 192.168.2.117 (192.168.2.117),
Dst: 192.168.2.113 (192.168.2.113)
Stream Control Transmission Protocol, Src Port: s1-control (36412),
Dst Port: s1-control (36412)
S1 Application Protocol
No.
    25 id-uplinkNASTransport, Uplink NAS transport(DTAP)
(SMS) CP-DATA (RP) RP-ACK (MS to Network)

Frame 25: 132 bytes on wire (1056 bits), 132 bytes captured
(1056 bits)
Linux cooked capture
```

```
Internet Protocol Version 4, Src: 192.168.2.117 (192.168.2.117),
Dst: 192.168.2.113 (192.168.2.113)
Stream Control Transmission Protocol, Src Port: s1-control (36412),
Dst Port: s1-control (36412)
S1 Application Protocol
No.
    26 id-downlinkNASTransport, Downlink NAS transport(DTAP)
(SMS) CP-ACK

Frame 26: 104 bytes on wire (832 bits), 104 bytes captured (832 bits)
Linux cooked capture
Internet Protocol Version 4, Src: 192.168.2.113 (192.168.2.113),
Dst: 192.168.2.117 (192.168.2.117)
Stream Control Transmission Protocol, Src Port: s1-control (36412),
Dst Port: s1-control (36412)
S1 Application Protocol

No.
    27 cmd=UnknownAnswer(8388646) flags=-P-- appl=MT_FORWARD_
SHORT_MSG(16777313) h2h=1e361f e2e=42c0a56c

Frame 27: 268 bytes on wire (2144 bits), 268 bytes captured
(2144 bits)
Linux cooked capture
Internet Protocol Version 4, Src: 192.168.2.113 (192.168.2.113),
Dst: 192.168.2.114 (192.168.2.114)
Stream Control Transmission Protocol, Src Port: diameter (3868),
Dst Port: diameter (3868)
Diameter Protocol
    Version: 0x01
    Length: 204
    Flags: 0x40
    Command Code: 8388646 MT_FORWARD_SHORT_MSG Ack
    ApplicationId: SGd (16777313)
    Hop-by-Hop Identifier: 0x001e361f
    End-to-End Identifier: 0x42c0a56c
    AVP: Session-Id(263) l=34 f=-M- val=smscHalys;1493279010;35571
    AVP: Vendor-Specific-Application-Id(260) l=32 f=-M-
    AVP: Result-Code(268) l=12 f=-M- val=DIAMETER_SUCCESS (2001)
    AVP: Auth-Session-State(277) l=12 f=-M- val=NO_STATE_
    MAINTAINED (1)
    AVP: Origin-Host(264) l=45 f=-M- val=mme.epc.mnc094.mcc208.
    3gppnetwork.org
    AVP: Origin-Realm(296) l=41 f=-M- val=epc.mnc094.mcc208.
    3gppnetwork.org
```

11.2.1 SMS-MO with diameter UE → SMSC (Chapter 4)

Figure 11.4 gives the high level PCAP traces of the sending of the SMS-MO to the SMSC with the report.

```
 5 id-uplinkNASTransport, Uplink NAS transport(DTAP) (SMS) CP-DATA (RP) RP-DATA (MS to Network)
 6 cmd=UnknownRequest(8388645) flags=RP-- appl=Unknown(16777313) h2h=273ab7d e2e=a83eada5
 7 SACK id-downlinkNASTransport, Downlink NAS transport(DTAP) (SMS) CP-ACK
 8 cmd=UnknownAnswer(8388645) flags=---- appl=Unknown(16777313) h2h=273ab7d e2e=a83eada5
 9 id-downlinkNASTransport, Downlink NAS transport(DTAP) (SMS) CP-DATA (RP) RP-ACK (Network to MS)
10 SACK id-uplinkNASTransport, Uplink NAS transport(DTAP) (SMS) CP-ACK
11 cmd=UnknownRequest(8388646) flags=R--- appl=Unknown(16777313) h2h=1e4335 e2e=42c0a574
12 id-downlinkNASTransport, Downlink NAS transport(DTAP) (SMS) CP-DATA (RP) RP-DATA (Network to MS)
13 SACK id-uplinkNASTransport, Uplink NAS transport(DTAP) (SMS) CP-ACK
14 id-uplinkNASTransport, Uplink NAS transport(DTAP) (SMS) CP-DATA (RP) RP-ACK (MS to Network)
15 id-downlinkNASTransport, Downlink NAS transport(DTAP) (SMS) CP-ACK
16 cmd=UnknownAnswer(8388646) flags=-P-- appl=Unknown(16777313) h2h=1e4335 e2e=42c0a574
```

Figure 11.4   FW-SM-MO call flow S1-AP eNodeB-MME, SGd MME-SMSC and GTP-C MME-S/PGW.

Trace details

```
No.
       5 id-uplinkNASTransport, Uplink NAS transport(DTAP) (SMS)
CP-DATA (RP) RP-DATA (MS to Network)

Frame 5: 156 bytes on wire (1248 bits), 156 bytes captured (1248 bits)
Linux cooked capture
Internet Protocol Version 4, Src: 192.168.2.117 (192.168.2.117), Dst:
192.168.2.113 (192.168.2.113)
Stream Control Transmission Protocol, Src Port: s1-control (36412),
Dst Port: s1-control (36412)
S1 Application Protocol
    S1AP-PDU: initiatingMessage (0)
        initiatingMessage
            procedureCode: id-uplinkNASTransport (13)
            criticality: ignore (1)
            value
                UplinkNASTransport
                    protocolIEs: 5 items
                        Item 0: id-MME-UE-S1AP-ID
                        Item 1: id-eNB-UE-S1AP-ID
                        Item 2: id-NAS-PDU
                            ProtocolIE-Field
                                id: id-NAS-PDU (26)
                                criticality: reject (0)
                                value
                                    NAS-PDU:
27b66383a10607631f39011c00040007918497903980f010...
                                    Non-Access-Stratum (NAS)PDU
                                    0010 .... = Security header type:
Integrity protected and ciphered (2)
                                    .... 0111 = Protocol
discriminator: EPS mobility management messages (0x07)
```

```
                              Message authentication code:
0xb66383a1
                              Sequence number: 6
                              0000 .... = Security header type:
Plain NAS message, not security protected (0)
                              .... 0111 = Protocol
discriminator: EPS mobility management messages (0x07)
                              NAS EPS Mobility Management
Message Type: Uplink NAS transport (0x63)
                              NAS message container
                                  Length: 31
                              NAS message container content:
39011c00040007918497903980f01021fc0b913379605701...
                                      GSM A-I/F DTAP - CP-DATA
                                          Protocol
                                          Discriminator:
                                          SMS messages
                                          DTAP Short Message
                                          Service Message
                                          Type: CP-DATA
                                          (0x01)
                                          CP-User Data
                                              Length: 28
                                              RPDU (not
                                              displayed)
                                      GSM A-I/F RP - RP-DATA
                                      (MS to Network)
                                          Message Type RP-DATA
                                          (MS to Network)
                                          RP-Message Reference
                                              RP-Message
                                              Reference:
                                              0x04 (4)
                                          RP-Originator Address
                                              Length: 0
                                          RP-Destination
                                          Address -
                                          (48790993080)
                                              Length: 7
                                              1... .... =
                                              Extension:
                                              No Extension
                                              .001 .... = Type
of number: International Number (0x01)
                                              .... 0001 =
Numbering plan identification: ISDN/Telephony Numbering (ITU-T Rec. E
    .164 / ITU-T Rec. E.163) (0x01)
                                              BCD Digits:
                                              48790993080
                                          RP-User Data
                                              Length: 16
                                              TPDU (not
                                              displayed)
```

```
                                        GSM SMS TPDU (GSM 03.40)
                                        SMS-SUBMIT
                                            0... .... = TP-RP:
TP Reply Path parameter is not set in this SMS SUBMIT/DELIVER
                                            .0.. .... = TP-UDHI:
The TP UD field contains only the short message
                                            ..1. .... = TP-SRR:
A status report is requested
                                            ...0 0... = TP-VPF:
TP-VP field not present (0)
                                            .... .0.. = TP-RD:
Instruct SC to accept duplicates
                                            .... ..01 = TP-MTI:
SMS-SUBMIT (1)
                                        TP-MR: 252
                                        TP-Destination-Address
                                        - (33970675102)
                                        TP-PID: 0
                                        TP-DCS: 0
                                            00.. .... = Coding
Group Bits: General Data Coding indication (0)
                                                Special case,
GSM 7 bit default alphabet
                                            TP-User-Data-
Length: (3) depends on Data-Coding-Scheme
                                        TP-User-Data
                                            SMS text:
                                            Ok2
              Item 3: id-EUTRAN-CGI
              Item 4: id-TAI

No.
    6 cmd=UnknownRequest(8388645) flags=RP-- appl=Unknown(16777313)
h2h=273ab7d e2e=a83eada5

Frame 6: 480 bytes on wire (3840 bits), 480 bytes captured (3840 bits)
Linux cooked capture
Internet Protocol Version 4, Src: 192.168.2.113 (192.168.2.113), Dst:
192.168.2.114 (192.168.2.114)
Stream Control Transmission Protocol, Src Port: 3871 (3871),
Dst Port: 3871 (3871)
Diameter Protocol
    Version: 0x01
    Length: 416
    Flags: 0xc0
    Command Code: 8388645 Unknown
    ApplicationId: Unknown (16777313)
    Hop-by-Hop Identifier: 0x0273ab7d
    End-to-End Identifier: 0xa83eada5
    [Answer In: 8]
    AVP: Session-Id(263) l=59 f=-M- val=mme.epc.mnc094.mcc208.
    3gppnetwork.org;1420130947;89
    AVP: Vendor-Specific-Application-Id(260) l=32 f=-M-
```

```
    AVP: Auth-Session-State(277) l=12 f=-M- val=NO_STATE_
    MAINTAINED (1)
    AVP: Origin-Host(264) l=45 f=-M- val=mme.epc.mnc094.mcc208.
    3gppnetwork.org
    AVP: Origin-Realm(296) l=41 f=-M- val=epc.mnc094.mcc208.
    3gppnetwork.org
    AVP: Destination-Host(293) l=46 f=-M- val=smsc.epc.mnc094.mcc208.
    3gppnetwork.org
    AVP: Destination-Realm(283) l=41 f=-M- val=epc.mnc094.mcc208.
    3gppnetwork.org
    AVP: Unknown(3300) l=23 f=VM- vnd=TGPP val=3438373930393933303830
    // SC-Address GT
    AVP: Unknown(3102) l=56 f=VM- vnd=TGPP
    val=0000000140000017323630303630313639323939937363600...
    AVP: Unknown(3301) l=28 f=VM- vnd=TGPP
    val=21fc0b913379605701f2000003cfb50c // SM-RP-UI

No.
     7 SACK id-downlinkNASTransport, Downlink NAS transport(DTAP)
(SMS) CP-ACK

Frame 7: 120 bytes on wire (960 bits), 120 bytes captured (960 bits)
Linux cooked capture
Internet Protocol Version 4, Src: 192.168.2.113 (192.168.2.113), Dst:
192.168.2.117 (192.168.2.117)
Stream Control Transmission Protocol, Src Port: s1-control (36412),
Dst Port: s1-control (36412)
S1 Application Protocol

No.
     8 cmd=UnknownAnswer(8388645) flags=----- appl=Unknown(16777313)
h2h=273ab7d e2e=a83eada5

Frame 8: 260 bytes on wire (2080 bits), 260 bytes captured (2080 bits)
Linux cooked capture
Internet Protocol Version 4, Src: 192.168.2.114 (192.168.2.114), Dst:
192.168.2.113 (192.168.2.113)
Stream Control Transmission Protocol, Src Port: 3871 (3871),
Dst Port: 3871 (3871)
Diameter Protocol
    Version: 0x01
    Length: 196
    Flags: 0x00
    Command Code: 8388645 Unknown
    ApplicationId: Unknown (16777313)
    Hop-by-Hop Identifier: 0x0273ab7d
    End-to-End Identifier: 0xa83eada5
    [Request In: 6]
    [Response Time: 0.036838000 seconds]
    AVP: Session-Id(263) l=59 f=-M- val=mme.epc.mnc094.mcc208.
    3gppnetwork.org;1420130947;89
    AVP: Auth-Session-State(277) l=12 f=-M- val=NO_STATE_
    MAINTAINED (1)
```

```
    AVP: Origin-Host(264) l=46 f=-M- val=smsc.epc.mnc094.mcc208.
    3gppnetwork.org
    AVP: Origin-Realm(296) l=41 f=-M- val=epc.mnc094.mcc208.
    3gppnetwork.org
    AVP: Result-Code(268) l=12 f=-M- val=DIAMETER_SUCCESS (2001)
```

```
No.
     9 id-downlinkNASTransport, Downlink NAS transport(DTAP) (SMS)
CP-DATA (RP) RP-ACK (Network to MS)
```

```
Frame 9: 108 bytes on wire (864 bits), 108 bytes captured (864 bits)
Linux cooked capture
Internet Protocol Version 4, Src: 192.168.2.113 (192.168.2.113),
Dst: 192.168.2.117 (192.168.2.117)
Stream Control Transmission Protocol, Src Port: s1-control (36412),
Dst Port: s1-control (36412)
S1 Application Protocol
```

```
No.
    10 SACK id-uplinkNASTransport, Uplink NAS transport(DTAP)
(SMS) CP-ACK
```

```
Frame 10: 140 bytes on wire (1120 bits), 140 bytes captured
(1120 bits)
Linux cooked capture
Internet Protocol Version 4, Src: 192.168.2.117 (192.168.2.117), Dst:
192.168.2.113 (192.168.2.113)
Stream Control Transmission Protocol, Src Port: s1-control (36412),
Dst Port: s1-control (36412)
S1 Application Protocol
```

```
SMS status report
```

```
No.
    11 cmd=UnknownRequest(8388646) flags=R--- appl=Unknown(16777313)
h2h=1e4335 e2e=42c0a574
```

```
Frame 11: 404 bytes on wire (3232 bits), 404 bytes captured
(3232 bits)
Linux cooked capture
Internet Protocol Version 4, Src: 192.168.2.114 (192.168.2.114), Dst:
192.168.2.113 (192.168.2.113)
Stream Control Transmission Protocol, Src Port: 3871 (3871), Dst Port:
3871 (3871)
Diameter Protocol
    Version: 0x01
    Length: 340
    Flags: 0x80
    Command Code: 8388646 Unknown
    ApplicationId: Unknown (16777313)
    Hop-by-Hop Identifier: 0x001e4335
    End-to-End Identifier: 0x42c0a574
    AVP: Session-Id(263) l=34 f=-M- val=smscHalys;1493285614;35579
```

```
    AVP: Auth-Session-State(277) l=12 f=-M- val=NO_STATE_
    MAINTAINED (1)
    AVP: Origin-Host(264) l=46 f=-M- val=smsc.epc.mnc094.mcc208.
    3gppnetwork.org
    AVP: Origin-Realm(296) l=41 f=-M- val=epc.mnc094.mcc208.
    3gppnetwork.org
    AVP: Destination-Host(293) l=45 f=-M- val=mme.epc.mnc094.mcc208.
    3gppnetwork.org
    AVP: Destination-Realm(283) l=41 f=-M- val=epc.mnc094.mcc208.
    3gppnetwork.org
    AVP: User-Name(1) l=23 f=-M- val=260060169299766
    AVP: Unknown(3300) l=23 f=VM- vnd=TGPP val=3333393730363730303037
    AVP: Unknown(3301) l=38 f=VM- vnd=TGPP val=06fc0b913379605701f27
    14072113303807140721133380...

No.
      12 id-downlinkNASTransport, Downlink NAS transport(DTAP) (SMS)
CP-DATA (RP) RP-DATA (Network to MS)

Frame 12: 144 bytes on wire (1152 bits), 144 bytes captured
(1152 bits)
Linux cooked capture
Internet Protocol Version 4, Src: 192.168.2.113 (192.168.2.113),
Dst: 192.168.2.117 (192.168.2.117)
Stream Control Transmission Protocol, Src Port: s1-control (36412),
Dst Port: s1-control (36412)
S1 Application Protocol
    S1AP-PDU: initiatingMessage (0)
        initiatingMessage
            procedureCode: id-downlinkNASTransport (11)
            criticality: ignore (1)
            value
                DownlinkNASTransport
                    protocolIEs: 3 items
                        Item 0: id-MME-UE-S1AP-ID
                        Item 1: id-eNB-UE-S1AP-ID
                        Item 2: id-NAS-PDU
                            ProtocolIE-Field
                                id: id-NAS-PDU (26)
                            criticality: reject (0)
                            value
                                NAS-PDU:
27f5d030ce08076229390126010a07913379600700f7001a...
                                    Non-Access-Stratum (NAS)PDU
                                        0010 .... = Security header type:
Integrity protected and ciphered (2)
                                        .... 0111 = Protocol
discriminator: EPS mobility management messages (0x07)
                                        Message authentication code:
                                        0xf5d030ce
                                        Sequence number: 8
                                        0000 .... = Security header type:
Plain NAS message, not security protected (0)
```

```
                                    .... 0111 = Protocol
discriminator: EPS mobility management messages (0x07)
                                  NAS EPS Mobility Management
Message Type: Downlink NAS transport (0x62)
                                  NAS message container
                                     Length: 41
                                     NAS message container content:
390126010a07913379600700f7001a06fc0b913379605701...
                                              GSM A-I/F DTAP -
                                              CP-DATA
                                                 Protocol
                                                 Discriminator:
                                                 SMS messages
                                                 DTAP Short Message
                                                 Service Message
                                                 Type: CP-DATA
                                                 (0x01)
                                                 CP-User Data
                                                    Length: 38
                                                    RPDU (not
                                                    displayed)
                                              GSM A-I/F RP - RP-DATA
     (Network to MS)
                                              GSM SMS TPDU
(GSM 03.40) SMS-STATUS REPORT
                                                 .0.. .... =
TP-UDHI: The TP UD field contains only the short message
                                                 ..0. .... =
TP-SRQ: SMS STATUS REPORT is the result of a SMS SUBMIT.
                                                 .... .1.. =
TP-MMS: No more messages are waiting for the MS in this SC
                                                 .... ..10 =
TP-MTI: SMS-STATUS REPORT (2)
                                                    TP-MR: 252
                                                    TP-Recipient-
                                                    Address -
                                                    (33970675102)
                                                       Length: 11
                                                       address
                                                       digits
                                                       1... .... : No
                                                       extension
                                                       .001 .... :
Type of number: (1) International
                                                       .... 0001 :
Numbering plan: (1) ISDN/telephone (E.164/E.163)
                                                    TP-RA Digits:
                                                    33970675102
                                                    TP-Service-Centre-
                                                    Time-Stamp
                                                    TP-Discharge-Time
                                                    TP-Status
                                                    TP-Parameter-
```

```
                                           Indicator

No.

    13 SACK id-uplinkNASTransport, Uplink NAS transport(DTAP)
(SMS) CP-ACK

Frame 13: 140 bytes on wire (1120 bits), 140 bytes captured
(1120 bits)
Linux cooked capture
Internet Protocol Version 4, Src: 192.168.2.117 (192.168.2.117),
Dst: 192.168.2.113 (192.168.2.113)
Stream Control Transmission Protocol, Src Port: s1-control (36412),
Dst Port: s1-control (36412)
S1 Application Protocol

No.

    14 id-uplinkNASTransport, Uplink NAS transport(DTAP) (SMS)
CP-DATA (RP) RP-ACK (MS to Network)

Frame 14: 128 bytes on wire (1024 bits), 128 bytes captured
(1024 bits)
Linux cooked capture
Internet Protocol Version 4, Src: 192.168.2.117 (192.168.2.117),
Dst: 192.168.2.113 (192.168.2.113)
Stream Control Transmission Protocol, Src Port: s1-control (36412),
Dst Port: s1-control (36412)
S1 Application Protocol
    S1AP-PDU: initiatingMessage (0)
        initiatingMessage
            procedureCode: id-uplinkNASTransport (13)
            criticality: ignore (1)
            value
                UplinkNASTransport
                    protocolIEs: 5 items
                        Item 0: id-MME-UE-S1AP-ID
                        Item 1: id-eNB-UE-S1AP-ID
                        Item 2: id-NAS-PDU
                            ProtocolIE-Field
                                id: id-NAS-PDU (26)
                                criticality: reject (0)
                                value
                                    NAS-PDU:
27efe083d809076305b90102020a
                                    Non-Access-Stratum (NAS)PDU
                                        0010 .... = Security header
type: Integrity protected and ciphered (2)
                                        .... 0111 = Protocol
discriminator: EPS mobility management messages (0x07)
                                        Message authentication code:
                                        0xefe083d8
                                        Sequence number: 9
```

```
                                        0000 .... = Security header
type: Plain NAS message, not security protected (0)
                                        .... 0111 = Protocol
discriminator: EPS mobility management messages (0x07)
                                        NAS EPS Mobility Management
Message Type: Uplink NAS transport (0x63)
                                        NAS message container
                                            Length: 5
                                            NAS message container
                                            content: b90102020a
                                                GSM A-I/F DTAP -
                                                CP-DATA
                                                    Protocol
                                                    Discriminator:
                                                    SMS messages
                                                        .... 1001 =
Protocol discriminator: SMS messages (0x09)
                                                        1... .... = TI
flag: allocated by receiver
                                                        .011 .... =
TIO: 3
                                                    DTAP Short Message
Service Message Type: CP-DATA (0x01)
                                                    CP-User Data
                                                        Length: 2
                                                        RPDU (not
                                                        displayed)
                                                GSM A-I/F RP - RP-ACK
                                                (MS to Network)
                Item 3: id-EUTRAN-CGI
                Item 4: id-TAI

No.
    15 id-downlinkNASTransport, Downlink NAS transport(DTAP)
(SMS) CP-ACK

Frame 15: 104 bytes on wire (832 bits), 104 bytes captured (832 bits)
Linux cooked capture
Internet Protocol Version 4, Src: 192.168.2.113 (192.168.2.113), Dst:
192.168.2.117 (192.168.2.117)
Stream Control Transmission Protocol, Src Port: s1-control (36412),
Dst Port: s1-control (36412)
S1 Application Protocol

No.
    16 cmd=UnknownAnswer(8388646) flags=-P-- appl=Unknown(16777313)
h2h=1e4335 e2e=42c0a574

Frame 16: 268 bytes on wire (2144 bits), 268 bytes captured
(2144 bits)
Linux cooked capture
Internet Protocol Version 4, Src: 192.168.2.113 (192.168.2.113), Dst:
192.168.2.114 (192.168.2.114)
```

```
Stream Control Transmission Protocol, Src Port: diameter (3868),
Dst Port: diameter (3868)
Diameter Protocol
    Version: 0x01
    Length: 204
    Flags: 0x40
    Command Code: 8388646 Unknown
    ApplicationId: Unknown (16777313)
    Hop-by-Hop Identifier: 0x001e4335
    End-to-End Identifier: 0x42c0a574
    AVP: Session-Id(263) l=34 f=-M- val=smscHalys;1493285614;35579
    AVP: Vendor-Specific-Application-Id(260) l=32 f=-M-
    AVP: Result-Code(268) l=12 f=-M- val=DIAMETER_SUCCESS (2001)
    AVP: Auth-Session-State(277) l=12 f=-M- val=NO_STATE_
    MAINTAINED (1)
    AVP: Origin-Host(264) l=45 f=-M- val=mme.epc.mnc094.mcc208.
    3gppnetwork.org
    AVP: Origin-Realm(296) l=41 f=-M- val=epc.mnc094.mcc208.
    3gppnetwork.org
```

11.3 Multicast: Traces M2AP, M3AP, Sgmb, and GTP V2 (Chapter 6)

In the traces below, the MCE and MME are on the same system with IP 192.168.2.113.

11.3.1 M2 SETUP request eNodeB → MCE

This is ASN1 (PER aligned) encoded Gdu.

11.3.2 M3 SETUP request MCE → MME

This is ASN1 (PER aligned) encoded. The MCE will receive the list of eNodeBs concerned by the multicast and will answer to the eNodeB which will open the MCCH multicast channels.

```
No. Time            Source           Destination    Protocol Length IMSI Info
1 11:12:45.550125 192.168.2.113 192.168.2.113 M3 Application Protocol
    88 M3 Setup Request

Frame 1: 88 bytes on wire (704 bits), 88 bytes captured (704 bits) on
interface 1
Linux cooked capture
Internet Protocol Version 4, Src: 192.168.2.113, Dst: 192.168.2.113
Stream Control Transmission Protocol, Src Port: 42817 (42817),
Dst Port: 36444 (36444)
M3 Application Protocol
```

```
M3AP-PDU: initiatingMessage (0)
    initiatingMessage
        procedureCode: id-m3Setup (7)
        criticality: reject (0)
        value
            M3SetupRequest
                protocolIEs: 2 items
                    Item 0: id-Global-MCE-ID
                        ProtocolIE-Field
                            id: id-Global-MCE-ID (18)
                            criticality: reject (0)
                            value
                                Global-MCE-ID
                                    pLMN-Identity: 02f849
                                    Mobile Country Code (MCC):
                                    France (208)
                                    Mobile Network Code (MNC):
                                    Halys (94)
                                    mCE-ID: 0001
                    Item 1: id-MBMSServiceAreaList
                        // MBMS Service Area List MCE
                        ProtocolIE-Field
                            id: id-MBMSServiceAreaList (20)
                            criticality: reject (0)
                            value
                                MBMSServiceAreaListItem: 1 item
                                Here only 1 Service Area
                                        Item 0
                                            MBMSServiceArea1:
                                            0001
```

11.3.3 M3 SETUP response MME → MCE

```
No. Time           Source          Destination    Protocol Length IMSI Info
2 11:12:45.550435 192.168.2.113 192.168.2.113 M3 Application Protocol
    72 M3 Setup Response

Frame 2: 72 bytes on wire (576 bits), 72 bytes captured (576 bits) on
interface 1
Linux cooked capture
Internet Protocol Version 4, Src: 192.168.2.113, Dst: 192.168.2.113
Stream Control Transmission Protocol, Src Port: 36444 (36444),
Dst Port: 42817 (42817)
M3 Application Protocol
    M3AP-PDU: successfulOutcome (1)
        successfulOutcome
            procedureCode: id-m3Setup (7)
            criticality: reject (0)
            value
                M3SetupResponse
                    protocolIEs: 0 items
```

11.3.4 M2 SETUP response MCE → NodeB

The MCE then responds to the eNodeB to ackowledge the end of the setup phase as in Figure 6.3.

11.3.5 Sgmb RAR BM-SC → MBMS GW

The BM-SC asks the MBMS GW to create a multicast bearer which will be identified by the couple TMGI x flow identifier below.

```
No. Time Source Destination Protocol Length EPS Bearer ID (EBI)
E.164 number (MSISDN) Charging id Identification F-TEID IPv4 Info
1 2017-09-01 16:21:01.686252 192.168.2.114 192.168.2.114 DIAMETER
600 0x0000 (0) cmd=Re-Auth Request(258) flags=R--- appl=3GPP
SGmb(16777292) h2h=4721fc e2e=1d |

Frame 1: 600 bytes on wire (4800 bits), 600 bytes captured (4800 bits)
on interface 0
Linux cooked capture
Internet Protocol Version 4, Src: 192.168.2.114, Dst: 192.168.2.114
Stream Control Transmission Protocol, Src Port: 4874 (4874),
Dst Port: 3874 (3874)
Diameter Protocol
    Version: 0x01
    Length: 536
    Flags: 0x80, Request
    Command Code: 258 Re-Auth
    ApplicationId: 3GPP SGmb (16777292)
    Hop-by-Hop Identifier: 0x004721fc
    End-to-End Identifier: 0x0000001d
    [Answer In: 3]
    AVP: Session-Id(263) l=62 f=-M- val=bmsc1.epc.mnc094.mcc208.
    3gppnetwork.org;1504275661;684
    AVP: Origin-Host(264) l=47 f=-M- val=bmsc1.epc.mnc094.mcc208.
    3gppnetwork.org
    AVP: Origin-Realm(296) l=41 f=-M- val=epc.mnc094.mcc208.
    3gppnetwork.org
    AVP: Destination-Realm(283) l=41 f=-M- val=epc.mnc094.mcc208.
    3gppnetwork.org
    AVP: Destination-Host(293) l=49 f=-M- val=mbmsgw1.epc.mnc094.
    mcc208.
    3gppnetwork.org
    AVP: Auth-Application-Id(258) l=12 f=-M- val=3GPP SGmb (16777292)
    AVP: Re-Auth-Request-Type(285) l=12 f=-M- val=AUTHORIZE_ONLY (0)
    AVP: MBMS-Access-Indicator(923) l=16 f=VM- vnd=TGPP
    val=E-UTRAN (1)
        AVP Code: 923 MBMS-Access-Indicator
        AVP Flags: 0xc0
        AVP Length: 16
        AVP Vendor Id: 3GPP (10415)
        MBMS-Access-Indicator: E-UTRAN (1)
```

```
    AVP: MBMS-StartStop-Indication(902) l=16 f=VM- vnd=TGPP
val=START (0)
        AVP Code: 902 MBMS-StartStop-Indication
        AVP Flags: 0xc0
        AVP Length: 16
        AVP Vendor Id: 3GPP (10415)
        MBMS-StartStop-Indication: START (0)
    AVP: MBMS-Service-Area(903) l=17 f=VM- vnd=TGPP val=0112345678
        AVP Code: 903 MBMS-Service-Area
        AVP Flags: 0xc0
        AVP Length: 17
        AVP Vendor Id: 3GPP (10415)
        MBMS-Service-Area: 0112345678
        Number of MBMS service area codes: 2
        MBMS service area code: 4660
        MBMS service area code: 22136
        Padding: 000000
    AVP: QoS-Information(1016) l=88 f=VM- vnd=TGPP
        AVP Code: 1016 QoS-Information
        AVP Flags: 0xc0
        AVP Length: 88
        AVP Vendor Id: 3GPP (10415)
        QoS-Information: 00000404
    c0000010000028af0000000400000203c0000010...
            AVP: QoS-Class-Identifier(1028) l=16 f=VM- vnd=TGPP
val=QCI_4 (4)
            AVP: Max-Requested-Bandwidth-DL(515) l=16 f=VM- vnd=TGPP
val=2000000
            AVP: Guaranteed-Bitrate-DL(1025) l=16 f=VM- vnd=TGPP
val=2000000
            AVP: Allocation-Retention-Priority(1034) l=28 f=V--
vnd=TGPP
    AVP: MBMS-Session-Duration(904) l=15 f=VM- vnd=TGPP val=001e00
        AVP Code: 904 MBMS-Session-Duration
        AVP Flags: 0xc0
        AVP Length: 15
        AVP Vendor Id: 3GPP (10415)
        MBMS-Session-Duration: 001e00
        .... .... .... .... .000 0000 = Estimated session duration
days: 0
        0000 0000 0001 1110 0... .... = Estimated session duration
seconds: 60
        Padding: 00
    AVP: TMGI(900) l=18 f=VM- vnd=TGPP val=12345602f849
        AVP Code: 900 TMGI
        AVP Flags: 0xc0
        AVP Length: 18
        AVP Vendor Id: 3GPP (10415)
        TMGI: 12345602f849
        Padding: 0000
    AVP: 3GPP-SGSN-Address(6) l=16 f=VM- vnd=TGPP val=192.168.2.113
        AVP Code: 6 3GPP-SGSN-Address
        AVP Flags: 0xc0
```

```
          AVP Length: 16
          AVP Vendor Id: 3GPP (10415)
          3GPP-SGSN-Address: c0a80271
     AVP: MBMS-Time-To-Data-Transfer(911) l=13 f=VM- vnd=TGPP val=09
          AVP Code: 911 MBMS-Time-To-Data-Transfer
          AVP Flags: 0xc0
          AVP Length: 13
          AVP Vendor Id: 3GPP (10415)
          MBMS-Time-To-Data-Transfer: 09
          Time to MBMS Data Transfer: 10
          Padding: 000000
     AVP: MBMS-User-Data-Mode-Indication(915) l=16 f=VM- vnd=TGPP
     val=Unicast (0)
          AVP Code: 915 MBMS-User-Data-Mode-Indication
          AVP Flags: 0xc0
          AVP Length: 16
          AVP Vendor Id: 3GPP (10415)
          MBMS-User-Data-Mode-Indication: Unicast (0)
     AVP: MBMS-Flow-Identifier(920) l=14 f=VM- vnd=TGPP val=1236
          AVP Code: 920 MBMS-Flow-Identifier
          AVP Flags: 0xc0
          AVP Length: 14
          AVP Vendor Id: 3GPP (10415)
          MBMS-Flow-Identifier: 1236
          Padding: 0000
```

11.3.6 GTPV2 MBMS session start request MBMS GW → MME

The MBMS GW sends a message to the MME containing the TMGI x flow
identifier received from the BM-SC and the IP Multicast that it allocates to the
new multicast bearer so that several UEs may receive it simultaneously (they
need to support IP Multicast, which is now available in all recent handsets).

```
No. Time              Source        Destination   Protocol Length IMSI Info
3 14:29:10.605856 192.168.2.114 192.168.2.113 GTPv2    149
    MBMS Session Start Request

Frame 3: 149 bytes on wire (1192 bits), 149 bytes captured (1192 bits)
on interface 1
Linux cooked capture
Internet Protocol Version 4, Src: 192.168.2.114, Dst: 192.168.2.113
User Datagram Protocol, Src Port: 2123, Dst Port: 2123
GPRS Tunneling Protocol V2
MBMS Session Start Request
    Flags: 0x48
        010. .... = Version: 2
        ...0 .... = Piggybacking flag (P): 0
        .... 1... = TEID flag (T): 1
    Message Type: MBMS Session Start Request (231)
    Message Length: 101
```

```
    Tunnel Endpoint Identifier: 0x00000000 (0)
    Sequence Number: 0x00000008 (8)
    Spare: 0
    Fully Qualified Tunnel Endpoint Identifier (F-TEID) : Sm MBMS
GW GTP-C interface, TEID/GRE Key: 0x50000000, IPv4 192.168.2.114
        IE Type: Fully Qualified Tunnel Endpoint Identifier
(F-TEID) (87)
        IE Length: 9
        0000 .... = CR flag: 0
        .... 0000 = Instance: 0
        1... .... = V4: IPv4 address present
        .0.. .... = V6: IPv6 address not present
        ..01 1000 = Interface Type: Sm MBMS GW GTP-C interface (24)
        TEID/GRE Key: 0x50000000
        F-TEID IPv4: 192.168.2.114
    Temporary Mobile Group Identity : 12345602f849
        IE Type: Temporary Mobile Group Identity (158)
        IE Length: 6
        0000 .... = CR flag: 0
        .... 0000 = Instance: 0
        MBMS Service ID: 123456
        Mobile Country Code (MCC): France (208)
        Mobile Network Code (MNC): Halys (94)
    MBMS Session Duration : 0 days 00:01:00 (DD days HH:MM:SS)
        IE Type: MBMS Session Duration (138)
        IE Length: 3
        0000 .... = CR flag: 0
        .... 0000 = Instance: 0
        .... .... .... .... .000 0000 = MBMS Session Duration
        (days): 0
        0000 0000 0001 1110 0... .... = MBMS Session Duration
        (seconds): 60
    MBMS Service Area : 1                     // MBMS Service Area List
                                                 MBMS GW
        IE Type: MBMS Service Area (139)
        IE Length: 3
        0000 .... = CR flag: 0
        .... 0000 = Instance: 0
        Number of MBMS Service Area codes: 1
        MBMS Service Area code (Service Area Identity): 1
    MBMS Flow Identifier : 1236
        IE Type: MBMS Flow Identifier (141)
        IE Length: 2
        0000 .... = CR flag: 0
        .... 0000 = Instance: 0
        MBMS Flow Identifier: 1236
    Bearer Level Quality of Service (Bearer QoS) :
        IE Type: Bearer Level Quality of Service (Bearer QoS) (80)
        IE Length: 22
        0000 .... = CR flag: 0
        .... 0000 = Instance: 0
        .... ...0 = PVI (Pre-emption Vulnerability): Enabled
        ..11 00.. = PL (Priority Level): 12
```

```
     .0.. .... = PCI (Pre-emption Capability): Enabled
     Label (QCI): 4
     Maximum Bit Rate For Uplink: 0
     Maximum Bit Rate For Downlink: 2000
     Guaranteed Bit Rate For Uplink: 0
     Guaranteed Bit Rate For Downlink: 2000
  MBMS IP Multicast Distribution : IPv4 Dist 232.1.100.1 IPv4 Src
192.168.2.114
     IE Type: MBMS IP Multicast Distribution (142)
     IE Length: 15
     0000 .... = CR flag: 0
     .... 0000 = Instance: 0
     Common Tunnel Endpoint Identifier: 80
     00.. .... = IP Address Type: 0
     ..00 0100 = IP Address Length: 4
     MBMS IP Multicast Distribution Address (IPv4): 232.1.100.1
     00.. .... = IP Address Type: 0
     ..00 0100 = IP Address Length: 4
     MBMS IP Multicast Source Address (IPv4): 192.168.2.114
     MBMS HC Indicator: Uncompressed header (0)
  MBMS Time to Data Transfer : 10 second(s)
     IE Type: MBMS Time to Data Transfer (153)
     IE Length: 1
     0000 .... = CR flag: 0
     .... 0000 = Instance: 0
     MBMS Time to Data Transfer: 10 second(s)
```

11.3.7 SGmb RAA MBMS GW → BM-SC

The RAA is returned just after the GTP v2 MBMS Start Session before the answer from the MME.

The MBMS GW has allocated and returns an AVP: MBMS-GW-UDP-Port which is used by the BM-SC to send multicast data with the SGi–mb interface to the specific MBMS bearer which has just been created. It also contains the AVP: MBMS-GGSN-Address for the eventual allocation of another MBMS GW IP for the SGI–mb data.

```
No. Time           Total Length Source          Destination
570 15:34:21.726921 596           192.168.2.114 192.168.2.114
Protocol Frame Number TEID Info
DIAMETER 570 SACK           cmd=Re-Auth Answer(258) flags=---- appl=
                            3GPP SGmb(16777292) h2h=5f0874 e2e=27 |

Frame 570: 612 bytes on wire (4896 bits), 612 bytes captured
(4896 bits) on interface 0
Linux cooked capture
Internet Protocol Version 4, Src: 192.168.2.114, Dst: 192.168.2.114
Stream Control Transmission Protocol, Src Port: 3874 (3874),
Dst Port: 4874 (4874)
```

```
Diameter Protocol
    Version: 0x01
    Length: 184
    Flags: 0x00
    Command Code: 258 Re-Auth
    ApplicationId: 3GPP SGmb (16777292)
    Hop-by-Hop Identifier: 0x005f0874
    End-to-End Identifier: 0x00000027
    [Request In: 568]
    [Response Time: 0.001178000 seconds]
    AVP: Session-Id(263) l=62 f=-M- val=bmsc1.epc.mnc094.mcc208.
    3gppnetwork.org;1506440061;724
        AVP Code: 263 Session-Id
        AVP Flags: 0x40
        AVP Length: 62
        Session-Id: bmsc1.epc.mnc094.mcc208.3gppnetwork.org;
        1506440061;724
        Padding: 0000
    AVP: Origin-Host(264) l=24 f=-M- val=mbms_gw.halys.fr
    AVP: Origin-Realm(296) l=16 f=-M- val=halys.fr
    AVP: Result-Code(268) l=12 f=-M- val=DIAMETER_SUCCESS (2001)
        AVP Code: 268 Result-Code
        AVP Flags: 0x40
        AVP Length: 12
        Result-Code: DIAMETER_SUCCESS (2001)
    AVP: MBMS-GGSN-Address(916) l=16 f=VM- vnd=TGPP val=c0a80272
        AVP Code: 916 MBMS-GGSN-Address
        AVP Flags: 0xc0
        AVP Length: 16
        AVP Vendor Id: 3GPP (10415)
        MBMS-GGSN-Address: c0a80272
    AVP: MBMS-User-Data-Mode-Indication(915) l=16 f=VM- vnd=TGPP
    val=Unicast (0)
        AVP Code: 915 MBMS-User-Data-Mode-Indication
        AVP Flags: 0xc0
        AVP Length: 16
        AVP Vendor Id: 3GPP (10415)
        MBMS-User-Data-Mode-Indication: Unicast (0)
    AVP: MBMS-GW-UDP-Port(927) l=14 f=V-- vnd=TGPP val=2162
        AVP Code: 927 MBMS-GW-UDP-Port
        AVP Flags: 0x80
        AVP Length: 14
        AVP Vendor Id: 3GPP (10415)
        MBMS-GW-UDP-Port: 2162
        UDP Port: 33
        Padding: 0000
```

11.3.8 M3 MBMS session start request MME → MCE

The MME sends an M3AP command to the MCE with the received TMGI in the GTPv2 Start Session Request, and the common areas between MBMS Service Area List MCE and MBMS Service Area List MBMS GW.

```
No. Time          Source          Destination   Protocol Length IMSI Info
4 11:26:44.799852 192.168.2.113 192.168.2.113 M3 Application Protocol
    140 MBMS Session Start Request

Frame 4: 140 bytes on wire (1120 bits), 140 bytes captured (1120 bits)
on interface 1
Linux cooked capture
Internet Protocol Version 4, Src: 192.168.2.113, Dst: 192.168.2.113
Stream Control Transmission Protocol, Src Port: 36444 (36444), Dst
    Port:
42817 (42817)
M3 Application Protocol
    M3AP-PDU: initiatingMessage (0)
        initiatingMessage
            procedureCode: id-mBMSsessionStart (0)
            criticality: reject (0)
            value
                MBMSSessionStartRequest
                    protocolIEs: 7 items
                        Item 0: id-MME-MBMS-M3AP-ID
                            ProtocolIE-Field
                                id: id-MME-MBMS-M3AP-ID (0)
                                criticality: reject (0)
                                value
                                    MME-MBMS-M3AP-ID: 3

                        Item 1: id-TMGI
                            ProtocolIE-Field
                                id: id-TMGI (2)
                                criticality: reject (0)
                                value
                                    TMGI
                                        pLMNidentity: 02f849
                                        Mobile Country Code (MCC):
                                        France (208)
                                        Mobile Network Code (MNC):
                                        Halys (94)
                                        serviceID: 123456
                        Item 2: id-MBMS-E-RAB-QoS-Parameters
                            ProtocolIE-Field
                                id: id-MBMS-E-RAB-QoS-Parameters (4)
                                criticality: reject (0)
                                value
                                    MBMS-E-RAB-QoS-Parameters
                                        qCI: 4
                                        gbrQosInformation
                                            mBMS-E-RAB-
```

```
                           MaximumBitrateDL: 2000000
                              mBMS-E-RAB-
                           GuaranteedBitrateDL: 2000000
Item 3: id-MBMS-Session-Duration
    ProtocolIE-Field
        id: id-MBMS-Session-Duration (5)
        criticality: reject (0)
        value
            MBMS-Session-Duration: 0f0000
0 days 02:08:00 (DD days HH:MM:SS)
            .... .... .... .... .000 0000 =
MBMS Session Duration (days): 0
            0000 1111 0000 0000 0... .... =
MBMS Session Duration (seconds): 7680
Item 4: id-MBMS-Service-Area
    ProtocolIE-Field
        id: id-MBMS-Service-Area (6)
        criticality: reject (0)
        value
            MBMS-Service-Area: 000001 1
            Number of MBMS Service Area
            codes: 1
            MBMS Service Area code
            (Service Area Identity): 1
Item 5: id-MinimumTimeToMBMSDataTransfer
    ProtocolIE-Field
        id: id-MinimumTimeToMBMSDataTransfer
        (16)
        criticality: reject (0)
        value
            MinimumTimeToMBMSDataTransfer:
            0a 11 second(s)
            MBMS Time to Data Transfer:
            11 second(s)
Item 6: id-TNL-Information // IP Multicast
    ProtocolIE-Field
        id: id-TNL-Information (7)
        criticality: reject (0)
        value
            TNL-Information
                iPMCAddress: e8016401
                IPAddress: 232.1.100.1
                iPSourceAddress: c0a80272
                IPAddress: 192.168.2.114
                gTP-DLTEID: 00000050
```

11.3.9 M3 MBMS session start response MCE → MME

After the M2 protocol response from the eNodeB, the MCE responds to the MME.

```
No. Time           Source         Destination    Protocol Length IMSI Info
5 14:29:10.607177 192.168.2.113 192.168.2.113 M3 Application
Protocol 100 SACK MBMS Session Start Response

Frame 5: 100 bytes on wire (800 bits), 100 bytes captured (800 bits)
on interface 1
Linux cooked capture
Internet Protocol Version 4, Src: 192.168.2.113, Dst: 192.168.2.113
Stream Control Transmission Protocol, Src Port: 57215 (57215),
Dst Port: 36444 (36444)
M3 Application Protocol
    M3AP-PDU: successfulOutcome (1)
        successfulOutcome
            procedureCode: id-mBMSsessionStart (0)
            criticality: reject (0)
            value
                MBMSSessionStartResponse
                    protocolIEs: 2 items
                        Item 0: id-MME-MBMS-M3AP-ID
                            ProtocolIE-Field
                                id: id-MME-MBMS-M3AP-ID (0)
                                criticality: reject (0)
                                value
                                    MME-MBMS-M3AP-ID: 1
                        Item 1: id-MCE-MBMS-M3AP-ID
                            ProtocolIE-Field
                                id: id-MCE-MBMS-M3AP-ID (1)
                                criticality: reject (0)
                                value
                                    MCE-MBMS-M3AP-ID: 2
```

11.3.10 GTPV2 MBMS session start response MME → MBMS GW

When the MME has the response from the MCE, it responds to the MBMS GW

```
No. Time           Source         Destination    Protocol
6 14:29:10.607580 192.168.2.113 192.168.2.114 GTPv2
Length IMSI Info
75              MBMS Session Start Response

Frame 6: 75 bytes on wire (600 bits), 75 bytes captured (600 bits) on
interface 1
Linux cooked capture
Internet Protocol Version 4, Src: 192.168.2.113, Dst: 192.168.2.114
```

```
User Datagram Protocol, Src Port: 2123, Dst Port: 2123
GPRS Tunneling Protocol V2
MBMS Session Start Response
    Flags: 0x48
        010. .... = Version: 2
        ...0 .... = Piggybacking flag (P): 0
        .... 1... = TEID flag (T): 1
    Message Type: MBMS Session Start Response (232)
    Message Length: 27
    Tunnel Endpoint Identifier: 0x50000000 (1342177280)
    Sequence Number: 0x00000008 (8)
    Spare: 0
    Cause : Request accepted (16)
        IE Type: Cause (2)
        IE Length: 2
        0000 .... = CR flag: 0
        .... 0000 = Instance: 0
        Cause: Request accepted (16)
        0000 0... = Spare bit(s): 0
        .... .0.. = PCE (PDN Connection IE Error): False
        .... ..0. = BCE (Bearer Context IE Error): False
        .... ...0 = CS (Cause Source): Originated by node sending the
        message
    Fully Qualified Tunnel Endpoint Identifier (F-TEID) : Sm MME GTP-C
    interface, TEID/GRE Key: 0x00000001, IPv4 192.168.2.113
        IE Type: Fully Qualified Tunnel Endpoint Identifier
        (F-TEID) (87)
        IE Length: 9
        0000 .... = CR flag: 0
        .... 0000 = Instance: 0
        1... .... = V4: IPv4 address present
        .0.. .... = V6: IPv6 address not present
        ..01 1010 = Interface Type: Sm MME GTP-C interface (26)
        TEID/GRE Key: 0x00000001
        F-TEID IPv4: 192.168.2.113
```

11.4 Traces IoT (Chapter 10)

All the Diameter traces below are conveniently captured at the DRA level of the EPC except the uplink NAS Transport (SMS-MT) and downlink NAS Transport (SMS-DELIVER-REPORT) which are captured at the S1–AP interface MME-eNodeB.

11.4.1 Subscriber information request (SIR) and answer (SIA) MTC-IWF→ HSS S6m/diameter

```
No. Time            Source        Destination   Protocol Length IMSI
1 11:25:16,695993 192.168.2.114 192.168.2.112 DIAMETER 390
Signalling Link Selector Info
    cmd=3GPP-Subscriber-Information Request(8388641) flags=R--- appl=3
    GPP S6m(16777310) h2h=311463 e2e=16 |

Frame 1: 390 bytes on wire (3120 bits), 390 bytes captured (3120 bits)
     on
interface 0
Ethernet II, Src: Vmware_58:44:63 (00:0c:29:58:44:63),
Dst: Vmware_2d:c1:4c (00:0c:29:2d:c1:4c)
Internet Protocol Version 4, Src: 192.168.2.114, Dst: 192.168.2.112
Stream Control Transmission Protocol, Src Port: 43157 (43157), Dst
    Port: 8000 (8000)
Diameter Protocol
    Version: 0x01
    Length: 328
    Flags: 0x80, Request
        1... .... = Request: Set
        .0.. .... = Proxyable: Not set
        ..0. .... = Error: Not set
        ...0 .... = T(Potentially re-transmitted message): Not set
        .... 0... = Reserved: Not set
        .... .0.. = Reserved: Not set
        .... ..0. = Reserved: Not set
        .... ...0 = Reserved: Not set
    Command Code: 8388641 3GPP-Subscriber-Information Request
    ApplicationId: 3GPP S6m (16777310)
    Hop-by-Hop Identifier: 0x00311463
    End-to-End Identifier: 0x00000016
    [Answer In: 2]
    AVP: Session-Id(263) l=61 f=-M- val=mtc1.epc.mnc094.mcc208.
    3gppnetwork.org;1509618316;694
        AVP Code: 263 Session-Id
        AVP Flags: 0x40
            0... .... = Vendor-Specific: Not set
            .1.. .... = Mandatory: Set
            ..0. .... = Protected: Not set
            ...0 .... = Reserved: Not set
            .... 0... = Reserved: Not set
            .... .0.. = Reserved: Not set
            .... ..0. = Reserved: Not set
            .... ...0 = Reserved: Not set
        AVP Length: 61
        Session-Id: mtc1.epc.mnc094.mcc208.3gppnetwork.org;
        1509618316;694
        Padding: 000000
    AVP: Auth-Session-State(277) l=12 f=-M- val=NO_STATE_
    MAINTAINED (1)
```

```
      AVP Code: 277 Auth-Session-State
      AVP Flags: 0x40
          0... .... = Vendor-Specific: Not set
          .1.. .... = Mandatory: Set
          ..0. .... = Protected: Not set
          ...0 .... = Reserved: Not set
          .... 0... = Reserved: Not set
          .... .0.. = Reserved: Not set
          .... ..0. = Reserved: Not set
          .... ...0 = Reserved: Not set
      AVP Length: 12
      Auth-Session-State: NO_STATE_MAINTAINED (1)
AVP: Origin-Host(264) l=46 f=-M- val=mtc1.epc.mnc094.mcc208.
3gppnetwork.org
      AVP Code: 264 Origin-Host
      AVP Flags: 0x40
          0... .... = Vendor-Specific: Not set
          .1.. .... = Mandatory: Set
          ..0. .... = Protected: Not set
          ...0 .... = Reserved: Not set
          .... 0... = Reserved: Not set
          .... .0.. = Reserved: Not set
          .... ..0. = Reserved: Not set
          .... ...0 = Reserved: Not set
      AVP Length: 46
      Origin-Host: mtc1.epc.mnc094.mcc208.3gppnetwork.org
      Padding: 0000
AVP: Origin-Realm(296) l=41 f=-M- val=epc.mnc094.mcc208.
3gppnetwork.org
      AVP Code: 296 Origin-Realm
      AVP Flags: 0x40
          0... .... = Vendor-Specific: Not set
          .1.. .... = Mandatory: Set
          ..0. .... = Protected: Not set
          ...0 .... = Reserved: Not set
          .... 0... = Reserved: Not set
          .... .0.. = Reserved: Not set
          .... ..0. = Reserved: Not set
          .... ...0 = Reserved: Not set
      AVP Length: 41
      Origin-Realm: epc.mnc094.mcc208.3gppnetwork.org
      Padding: 000000
AVP: Destination-Realm(283) l=41 f=-M- val=epc.mnc094.mcc208.
3gppnetwork.org
      AVP Code: 283 Destination-Realm
      AVP Flags: 0x40
          0... .... = Vendor-Specific: Not set
          .1.. .... = Mandatory: Set
          ..0. .... = Protected: Not set
          ...0 .... = Reserved: Not set
          .... 0... = Reserved: Not set
          .... .0.. = Reserved: Not set
          .... ..0. = Reserved: Not set
```

```
              .... ...0 = Reserved: Not set
      AVP Length: 41
      Destination-Realm: epc.mnc094.mcc208.3gppnetwork.org
      Padding: 000000
AVP: User-Identifier(3102) l=32 f=VM- vnd=TGPP
      AVP Code: 3102 User-Identifier
      AVP Flags: 0xc0
          1... .... = Vendor-Specific: Set
          .1.. .... = Mandatory: Set
          ..0. .... = Protected: Not set
          ...0 .... = Reserved: Not set
          .... 0... = Reserved: Not set
          .... .0.. = Reserved: Not set
          .... ..0. = Reserved: Not set
          .... ...0 = Reserved: Not set
      AVP Length: 32
      AVP Vendor Id: 3GPP (10415)
      User-Identifier: 000002bdc0000012000028af3357771910f40000
          AVP: MSISDN(701) l=18 f=VM- vnd=TGPP val=3357771910f4
              AVP Code: 701 MSISDN
              AVP Flags: 0xc0
                  1... .... = Vendor-Specific: Set
                  .1.. .... = Mandatory: Set
                  ..0. .... = Protected: Not set
                  ...0 .... = Reserved: Not set
                  .... 0... = Reserved: Not set
                  .... .0.. = Reserved: Not set
                  .... ..0. = Reserved: Not set
                  .... ...0 = Reserved: Not set
              AVP Length: 18
              AVP Vendor Id: 3GPP (10415)
              MSISDN: 3357771910f4
              E.164 number (MSISDN): 33757791014
                  Country Code: France (33)
              Padding: 0000
AVP: SIR-Flags(3110) l=16 f=VM- vnd=TGPP val=1
      AVP Code: 3110 SIR-Flags
      AVP Flags: 0xc0
          1... .... = Vendor-Specific: Set
          .1.. .... = Mandatory: Set
          ..0. .... = Protected: Not set
          ...0 .... = Reserved: Not set
          .... 0... = Reserved: Not set
          .... .0.. = Reserved: Not set
          .... ..0. = Reserved: Not set
          .... ...0 = Reserved: Not set
      AVP Length: 16
      AVP Vendor Id: 3GPP (10415)
      SIR-Flags: 1
AVP: Destination-Host(293) l=46 f=-M- val=hss1.epc.mnc094.mcc208.
3gppnetwork.org
      AVP Code: 293 Destination-Host
      AVP Flags: 0x40
```

```
            0... .... = Vendor-Specific: Not set
            .1.. .... = Mandatory: Set
            ..0. .... = Protected: Not set
            ...0 .... = Reserved: Not set
            .... 0... = Reserved: Not set
            .... .0.. = Reserved: Not set
            .... ..0. = Reserved: Not set
            .... ...0 = Reserved: Not set
        AVP Length: 46
        Destination-Host: hss1.epc.mnc094.mcc208.3gppnetwork.org
        Padding: 0000

No. Time Source Destination Protocol Length IMSI Signalling Link
    Selector Info
2 11:25:16,700818 192.168.2.112 192.168.2.114 DIAMETER 334 SACK
cmd=3GPP-Subscriber-Information Answer(8388641) flags=---- appl=3GPP
S6m(16777310) h2h=311463 e2e=16 |

Frame 2: 334 bytes on wire (2672 bits), 334 bytes captured (2672 bits)
    on interface 0
Ethernet II, Src: Vmware_2d:c1:4c (00:0c:29:2d:c1:4c), Dst:
    Vmware_58:44:63 (00:0c:29:58:44:63)
Internet Protocol Version 4, Src: 192.168.2.112, Dst: 192.168.2.114
Stream Control Transmission Protocol, Src Port: 8000 (8000),
Dst Port: 43157 (43157)
Diameter Protocol
    Version: 0x01
    Length: 256
    Flags: 0x00
        0... .... = Request: Not set
        .0.. .... = Proxyable: Not set
        ..0. .... = Error: Not set
        ...0 .... = T(Potentially re-transmitted message): Not set
        .... 0... = Reserved: Not set
        .... .0.. = Reserved: Not set
        .... ..0. = Reserved: Not set
        .... ...0 = Reserved: Not set
    Command Code: 8388641 3GPP-Subscriber-Information Answer
    ApplicationId: 3GPP S6m (16777310)
    Hop-by-Hop Identifier: 0x00311463
    End-to-End Identifier: 0x00000016
    [Request In: 1]
    [Response Time: 0.004825000 seconds]
    AVP: Session-Id(263) l=61 f=-M- val=mtc1.epc.mnc094.mcc208.
    3gppnetwork.org;1509618316;694
        AVP Code: 263 Session-Id
        AVP Flags: 0x40
            0... .... = Vendor-Specific: Not set
            .1.. .... = Mandatory: Set
            ..0. .... = Protected: Not set
            ...0 .... = Reserved: Not set
            .... 0... = Reserved: Not set
            .... .0.. = Reserved: Not set
```

```
            .... ..0. = Reserved: Not set
            .... ...0 = Reserved: Not set
      AVP Length: 61
      Session-Id: mtc1.epc.mnc094.mcc208.3gppnetwork.org;
      1509618316;694
      Padding: 000000
AVP: Auth-Session-State(277) l=12 f=-M- val=NO_STATE_
MAINTAINED (1)
      AVP Code: 277 Auth-Session-State
      AVP Flags: 0x40
          0... .... = Vendor-Specific: Not set
          .1.. .... = Mandatory: Set
          ..0. .... = Protected: Not set
          ...0 .... = Reserved: Not set
          .... 0... = Reserved: Not set
          .... .0.. = Reserved: Not set
          .... ..0. = Reserved: Not set
          .... ...0 = Reserved: Not set
      AVP Length: 12
      Auth-Session-State: NO_STATE_MAINTAINED (1)
AVP: Origin-Host(264) l=48 f=-M- val=hlrhss.epc.mnc094.mcc208.
3gppnetwork.org
      AVP Code: 264 Origin-Host
      AVP Flags: 0x40
          0... .... = Vendor-Specific: Not set
          .1.. .... = Mandatory: Set
          ..0. .... = Protected: Not set
          ...0 .... = Reserved: Not set
          .... 0... = Reserved: Not set
          .... .0.. = Reserved: Not set
          .... ..0. = Reserved: Not set
          .... ...0 = Reserved: Not set
      AVP Length: 48
      Origin-Host: hlrhss.epc.mnc094.mcc208.3gppnetwork.org
AVP: Origin-Realm(296) l=41 f=-M- val=epc.mnc094.mcc208.
3gppnetwork.org
      AVP Code: 296 Origin-Realm
      AVP Flags: 0x40
          0... .... = Vendor-Specific: Not set
          .1.. .... = Mandatory: Set
          ..0. .... = Protected: Not set
          ...0 .... = Reserved: Not set
          .... 0... = Reserved: Not set
          .... .0.. = Reserved: Not set
          .... ..0. = Reserved: Not set
          .... ...0 = Reserved: Not set
      AVP Length: 41
      Origin-Realm: epc.mnc094.mcc208.3gppnetwork.org
      Padding: 000000
AVP: Result-Code(268) l=12 f=-M- val=DIAMETER_SUCCESS (2001)
      AVP Code: 268 Result-Code
      AVP Flags: 0x40
          0... .... = Vendor-Specific: Not set
```

```
                .1.. .... = Mandatory: Set
                ..0. .... = Protected: Not set
                ...0 .... = Reserved: Not set
                .... 0... = Reserved: Not set
                .... .0.. = Reserved: Not set
                .... ..0. = Reserved: Not set
                .... ...0 = Reserved: Not set
        AVP Length: 12
        Result-Code: DIAMETER_SUCCESS (2001)
    AVP: User-Identifier(3102) l=56 f=VM- vnd=TGPP
        AVP Code: 3102 User-Identifier
        AVP Flags: 0xc0
                1... .... = Vendor-Specific: Set
                .1.. .... = Mandatory: Set
                ..0. .... = Protected: Not set
                ...0 .... = Reserved: Not set
                .... 0... = Reserved: Not set
                .... .0.. = Reserved: Not set
                .... ..0. = Reserved: Not set
                .... ...0 = Reserved: Not set
        AVP Length: 56
        AVP Vendor Id: 3GPP (10415)
        User-Identifier:
00000001400000173230383934303030303030313031340 0...
            AVP: User-Name(1) l=23 f=-M- val=208940000001014
                AVP Code: 1 User-Name
                AVP Flags: 0x40
                        0... .... = Vendor-Specific: Not set
                        .1.. .... = Mandatory: Set
                        ..0. .... = Protected: Not set
                        ...0 .... = Reserved: Not set
                        .... 0... = Reserved: Not set
                        .... .0.. = Reserved: Not set
                        .... ..0. = Reserved: Not set
                        .... ...0 = Reserved: Not set
                AVP Length: 23
                User-Name: 208940000001014
                Padding: 00
            AVP: MSISDN(701) l=18 f=VM- vnd=TGPP val=3357771910f4
                AVP Code: 701 MSISDN
                AVP Flags: 0xc0
                        1... .... = Vendor-Specific: Set
                        .1.. .... = Mandatory: Set
                        ..0. .... = Protected: Not set
                        ...0 .... = Reserved: Not set
                        .... 0... = Reserved: Not set
                        .... .0.. = Reserved: Not set
                        .... ..0. = Reserved: Not set
                        .... ...0 = Reserved: Not set
                AVP Length: 18
```

```
AVP Vendor Id: 3GPP (10415)
MSISDN: 3357771910f4
E.164 number (MSISDN): 33757791014
    Country Code: France (33)
Padding: 0000
```

11.4.2 Devices trigger request (DTR) and answer (DTA) MTC-IWF→ SMSC T4/diameter

```
No. Time          Source        Destination   Protocol Length
3 11:25:49,215946 192.168.2.114 192.168.2.116 DIAMETER 526
IMSI Signalling Link Selector Info
    cmd=3GPP-Device-Trigger Request(8388643) flags=R--- appl=3GPP T4
    (16777311) h2h=311478 e2e=17 |

Frame 3: 526 bytes on wire (4208 bits), 526 bytes captured (4208 bits)
on interface 0
Ethernet II, Src: Vmware_58:44:63 (00:0c:29:58:44:63), Dst:
Vmware_ea:40:0f (00:0c:29:ea:40:0f)
Internet Protocol Version 4, Src: 192.168.2.114, Dst: 192.168.2.116
Stream Control Transmission Protocol, Src Port: 3873 (3873),
Dst Port: 3873 (3873)
Diameter Protocol
    Version: 0x01
    Length: 464
    Flags: 0x80, Request
        1... .... = Request: Set
        .0.. .... = Proxyable: Not set
        ..0. .... = Error: Not set
        ...0 .... = T(Potentially re-transmitted message): Not set
        .... 0... = Reserved: Not set
        .... .0.. = Reserved: Not set
        .... ..0. = Reserved: Not set
        .... ...0 = Reserved: Not set
    Command Code: 8388643 3GPP-Device-Trigger Request
    ApplicationId: 3GPP T4 (16777311)
    Hop-by-Hop Identifier: 0x00311478
    End-to-End Identifier: 0x00000017
    [Answer In: 4]
    AVP: Session-Id(263) l=61 f=-M- val=mtc1.epc.mnc094.mcc208.
3gppnetwork.org;1509618349;214
        AVP Code: 263 Session-Id
        AVP Flags: 0x40
            0... .... = Vendor-Specific: Not set
            .1.. .... = Mandatory: Set
            ..0. .... = Protected: Not set
            ...0 .... = Reserved: Not set
            .... 0... = Reserved: Not set
            .... .0.. = Reserved: Not set
            .... ..0. = Reserved: Not set
```

```
            .... ...0 = Reserved: Not set
      AVP Length: 61
      Session-Id: mtc1.epc.mnc094.mcc208.3gppnetwork.org;
      1509618349;214
      Padding: 000000
AVP: Auth-Session-State(277) l=12 f=-M- val=NO_STATE_
MAINTAINED (1)
      AVP Code: 277 Auth-Session-State
      AVP Flags: 0x40
          0... .... = Vendor-Specific: Not set
          .1.. .... = Mandatory: Set
          ..0. .... = Protected: Not set
          ...0 .... = Reserved: Not set
          .... 0... = Reserved: Not set
          .... .0.. = Reserved: Not set
          .... ..0. = Reserved: Not set
          .... ...0 = Reserved: Not set
      AVP Length: 12
      Auth-Session-State: NO_STATE_MAINTAINED (1)
      AVP: Origin-Host(264) l=46 f=-M- val=mtc1.epc.mnc094.mcc208.
      3gppnetwork.org
          AVP Code: 264 Origin-Host
          AVP Flags: 0x40
              0... .... = Vendor-Specific: Not set
              .1.. .... = Mandatory: Set
              ..0. .... = Protected: Not set
              ...0 .... = Reserved: Not set
              .... 0... = Reserved: Not set
              .... .0.. = Reserved: Not set
              .... ..0. = Reserved: Not set
              .... ...0 = Reserved: Not set
          AVP Length: 46
          Origin-Host: mtc1.epc.mnc094.mcc208.3gppnetwork.org
          Padding: 0000
      AVP: Origin-Realm(296) l=41 f=-M- val=epc.mnc094.mcc208.
      3gppnetwork.org
          AVP Code: 296 Origin-Realm
          AVP Flags: 0x40
              0... .... = Vendor-Specific: Not set
              .1.. .... = Mandatory: Set
              ..0. .... = Protected: Not set
              ...0 .... = Reserved: Not set
              .... 0... = Reserved: Not set
              .... .0.. = Reserved: Not set
              .... ..0. = Reserved: Not set
              .... ...0 = Reserved: Not set
          AVP Length: 41
          Origin-Realm: epc.mnc094.mcc208.3gppnetwork.org
          Padding: 000000
          AVP: Destination-Realm(283) l=41 f=-M- val=epc.mnc094.
          mcc208.3gppnetwork.org
              AVP Code: 283 Destination-Realm
              AVP Flags: 0x40
```

```
                      0... .... = Vendor-Specific: Not set
                      .1.. .... = Mandatory: Set
                      ..0. .... = Protected: Not set
                      ...0 .... = Reserved: Not set
                      .... 0... = Reserved: Not set
                      .... .0.. = Reserved: Not set
                      .... ..0. = Reserved: Not set
                      .... ...0 = Reserved: Not set
                  AVP Length: 41
                  Destination-Realm: epc.mnc094.mcc208.3gppnetwork.org
                  Padding: 000000
              AVP: User-Identifier(3102) l=36 f=VM- vnd=TGPP
                  AVP Code: 3102 User-Identifier
                  AVP Flags: 0xc0
                      1... .... = Vendor-Specific: Set
                      .1.. .... = Mandatory: Set
                      ..0. .... = Protected: Not set
                      ...0 .... = Reserved: Not set
                      .... 0... = Reserved: Not set
                      .... .0.. = Reserved: Not set
                      .... ..0. = Reserved: Not set
                      .... ...0 = Reserved: Not set
                  AVP Length: 36
                  AVP Vendor Id: 3GPP (10415)
                  User-Identifier:
                  00000001400000173230383934303030303030313031340 0
                      AVP: User-Name(1) l=23 f=-M- val=208940000001014
                          AVP Code: 1 User-Name
                          AVP Flags: 0x40
                              0... .... = Vendor-Specific: Not set
                              .1.. .... = Mandatory: Set
                              ..0. .... = Protected: Not set
                              ...0 .... = Reserved: Not set
                              .... 0... = Reserved: Not set
                              .... .0.. = Reserved: Not set
                              .... ..0. = Reserved: Not set
                              .... ...0 = Reserved: Not set
                          AVP Length: 23
                          User-Name: 208940000001014
                          Padding: 00
  AVP: SM-RP-SMEA(3309) l=18 f=VM- vnd=TGPP val=3379600710f3
      AVP Code: 3309 SM-RP-SMEA
      AVP Flags: 0xc0
          1... .... = Vendor-Specific: Set
          .1.. .... = Mandatory: Set
          ..0. .... = Protected: Not set
          ...0 .... = Reserved: Not set
          .... 0... = Reserved: Not set
          .... .0.. = Reserved: Not set
          .... ..0. = Reserved: Not set
          .... ...0 = Reserved: Not set
      AVP Length: 18
      AVP Vendor Id: 3GPP (10415)
```

```
     SM-RP-SMEA: 3379600710f3
     Padding: 0000
AVP: Payload(3004) l=13 f=VM- vnd=TGPP val=00
     AVP Code: 3004 Payload
     AVP Flags: 0xc0
          1... .... = Vendor-Specific: Set
          .1.. .... = Mandatory: Set
          ..0. .... = Protected: Not set
          ...0 .... = Reserved: Not set
          .... 0... = Reserved: Not set
          .... .0.. = Reserved: Not set
          .... ..0. = Reserved: Not set
          .... ...0 = Reserved: Not set
     AVP Length: 13
     AVP Vendor Id: 3GPP (10415)
     Payload: 00
     Padding: 000000
AVP: Destination-Host(293) l=47 f=-M- val=smsc1.epc.mnc094.mcc208.
3gppnetwork.org
     AVP Code: 293 Destination-Host
     AVP Flags: 0x40
          0... .... = Vendor-Specific: Not set
          .1.. .... = Mandatory: Set
          ..0. .... = Protected: Not set
          ...0 .... = Reserved: Not set
          .... 0... = Reserved: Not set
          .... .0.. = Reserved: Not set
          .... ..0. = Reserved: Not set
          .... ...0 = Reserved: Not set
     AVP Length: 47
     Destination-Host: smsc1.epc.mnc094.mcc208.3gppnetwork.org
     Padding: 00
AVP: Serving-Node(2401) l=112 f=VM- vnd=TGPP
     AVP Code: 2401 Serving-Node
     AVP Flags: 0xc0
          1... .... = Vendor-Specific: Set
          .1.. .... = Mandatory: Set
          ..0. .... = Protected: Not set
          ...0 .... = Reserved: Not set
          .... 0... = Reserved: Not set
          .... .0.. = Reserved: Not set
          .... ..0. = Reserved: Not set
          .... ...0 = Reserved: Not set
     AVP Length: 112
     AVP Vendor Id: 3GPP (10415)
     Serving-Node: 00000962c0000032000028af6d6d65312e6570632e6d6e
     63...
          AVP: MME-Name(2402) l=50 f=VM- vnd=TGPP val=mme1.epc.
          mnc094.mcc208.3gppnetwork.org
               AVP Code: 2402 MME-Name
               AVP Flags: 0xc0
                    1... .... = Vendor-Specific: Set
                    .1.. .... = Mandatory: Set
```

```
                            ..0. .... = Protected: Not set
                            ...0 .... = Reserved: Not set
                            .... 0... = Reserved: Not set
                            .... .0.. = Reserved: Not set
                            .... ..0. = Reserved: Not set
                            .... ...0 = Reserved: Not set
                    AVP Length: 50
                    AVP Vendor Id: 3GPP (10415)
                    MME-Name: mme1.epc.mnc094.mcc208.3gppnetwork.org
                    Padding: 0000
                AVP: MME-Realm(2408) l=45 f=V-- vnd=TGPP val=epc.mnc094.
                mcc208.3gppnetwork.org
                    AVP Code: 2408 MME-Realm
                    AVP Flags: 0x80
                            1... .... = Vendor-Specific: Set
                            .0.. .... = Mandatory: Not set
                            ..0. .... = Protected: Not set
                            ...0 .... = Reserved: Not set
                            .... 0... = Reserved: Not set
                            .... .0.. = Reserved: Not set
                            .... ..0. = Reserved: Not set
                            .... ...0 = Reserved: Not set
                    AVP Length: 45
                    AVP Vendor Id: 3GPP (10415)
                    MME-Realm: epc.mnc094.mcc208.3gppnetwork.org
                    Padding: 000000No. Time Source Destination Protocol
Length IMSI Signalling Link Selector Info
4 11:25:49,226539 192.168.2.116 192.168.2.114 DIAMETER 278 SACK
cmd=3GPP-Device-Trigger Answer(8388643) flags=---- appl=3GPP
T4(16777311) h2h=311478 e2e=17 |

Frame 4: 278 bytes on wire (2224 bits), 278 bytes captured (2224 bits)
on interface 0
Ethernet II, Src: Vmware_ea:40:0f (00:0c:29:ea:40:0f),
Dst: Vmware_58:44:63 (00:0c:29:58:44:63)
Internet Protocol Version 4, Src: 192.168.2.116, Dst: 192.168.2.114
Stream Control Transmission Protocol, Src Port: 3873 (3873),
Dst Port: 3873 (3873)
Diameter Protocol
    Version: 0x01
    Length: 200
    Flags: 0x00
        0... .... = Request: Not set
        .0.. .... = Proxyable: Not set
        ..0. .... = Error: Not set
        ...0 .... = T(Potentially re-transmitted message): Not set
        .... 0... = Reserved: Not set
        .... .0.. = Reserved: Not set
        .... ..0. = Reserved: Not set
        .... ...0 = Reserved: Not set
    Command Code: 8388643 3GPP-Device-Trigger Answer
    ApplicationId: 3GPP T4 (16777311)
    Hop-by-Hop Identifier: 0x00311478
```

```
End-to-End Identifier: 0x00000017
[Request In: 3]
[Response Time: 0.010593000 seconds]
AVP: Session-Id(263) l=61 f=-M- val=mtc1.epc.mnc094.mcc208.
3gppnetwork.org;1509618349;214
    AVP Code: 263 Session-Id
    AVP Flags: 0x40
        0... .... = Vendor-Specific: Not set
        .1.. .... = Mandatory: Set
        ..0. .... = Protected: Not set
        ...0 .... = Reserved: Not set
        .... 0... = Reserved: Not set
        .... .0.. = Reserved: Not set
        .... ..0. = Reserved: Not set
        .... ...0 = Reserved: Not set
    AVP Length: 61
    Session-Id: mtc1.epc.mnc094.mcc208.3gppnetwork.org
;1509618349;214
    Padding: 000000
AVP: Auth-Session-State(277) l=12 f=-M- val=NO_STATE_
MAINTAINED (1)
    AVP Code: 277 Auth-Session-State
    AVP Flags: 0x40
        0... .... = Vendor-Specific: Not set
        .1.. .... = Mandatory: Set
        ..0. .... = Protected: Not set
        ...0 .... = Reserved: Not set
        .... 0... = Reserved: Not set
        .... .0.. = Reserved: Not set
        .... ..0. = Reserved: Not set
        .... ...0 = Reserved: Not set
    AVP Length: 12
    Auth-Session-State: NO_STATE_MAINTAINED (1)
AVP: Origin-Host(264) l=47 f=-M- val=smsc1.epc.mnc094.mcc208.
3gppnetwork.org
    AVP Code: 264 Origin-Host
    AVP Flags: 0x40
        0... .... = Vendor-Specific: Not set
        .1.. .... = Mandatory: Set
        ..0. .... = Protected: Not set
        ...0 .... = Reserved: Not set
        .... 0... = Reserved: Not set
        .... .0.. = Reserved: Not set
        .... ..0. = Reserved: Not set
        .... ...0 = Reserved: Not set
    AVP Length: 47
    Origin-Host: smsc1.epc.mnc094.mcc208.3gppnetwork.org
    Padding: 00
AVP: Origin-Realm(296) l=41 f=-M- val=epc.mnc094.mcc208.
3gppnetwork.org
    AVP Code: 296 Origin-Realm
    AVP Flags: 0x40
        0... .... = Vendor-Specific: Not set
```

```
             .1.. .... = Mandatory: Set
             ..0. .... = Protected: Not set
             ...0 .... = Reserved: Not set
             .... 0... = Reserved: Not set
             ... .0.. = Reserved: Not set
             .... ..0. = Reserved: Not set
             .... ...0 = Reserved: Not set
      AVP Length: 41
      Origin-Realm: epc.mnc094.mcc208.3gppnetwork.org
      Padding: 000000
  AVP: Result-Code(268) l=12 f=-M- val=DIAMETER_SUCCESS (2001)
      AVP Code: 268 Result-Code
      AVP Flags: 0x40
             0... .... = Vendor-Specific: Not set
             .1.. .... = Mandatory: Set
             ..0. .... = Protected: Not set
             ...0 .... = Reserved: Not set
             .... 0... = Reserved: Not set
             .... .0.. = Reserved: Not set
             .... ..0. = Reserved: Not set
             .... ...0 = Reserved: Not set
      AVP Length: 12
      Result-Code: DIAMETER_SUCCESS (2001)
```

11.4.3 Transfer forward request (TFR) and answer (TFA) SMSC → MME Sgd/diameter (SMS-MT)

Wireshark does not fully decode the TFR and TFA, and hence we also give the trace at the S1-AP level.

```
No. Time              Source          Destination    Protocol Length IMSI
5   11:25:49,240324 192.168.2.116 192.168.2.114 DIAMETER 394
Signalling Link Selector Info
    cmd=Unknown Request(8388646) flags=R--- appl=3GPP SGd(16777313)
    h2h=5 e2e=4ec62f38 |

Frame 5: 394 bytes on wire (3152 bits), 394 bytes captured (3152 bits)
on interface 0
Ethernet II, Src: Vmware_ea:40:0f (00:0c:29:ea:40:0f), Dst:
Vmware_58:44:63 (00:0c:29:58:44:63)
Internet Protocol Version 4, Src: 192.168.2.116, Dst: 192.168.2.114
Stream Control Transmission Protocol, Src Port: 3873 (3873),
Dst Port: 3873 (3873) Diameter Protocol
    Version: 0x01
    Length: 332
    Flags: 0x80, Request
        1... .... = Request: Set
        .0.. .... = Proxyable: Not set
        ..0. .... = Error: Not set
        ...0 .... = T(Potentially re-transmitted message): Not set
        .... 0... = Reserved: Not set
```

```
        .... .0.. = Reserved: Not set
        .... ..0. = Reserved: Not set
        .... ...0 = Reserved: Not set
    Command Code: 8388646 Unknown Transfer Forward Request
        Unknown command, if you know what this is you can add it to
        dictionary.xml
            [Expert Info (Warning/Undecoded): Unknown command, if you
know what this is you can add it to dictionary.xml]
                [Unknown command, if you know what this is you can add
it to dictionary.xml]
                [Severity level: Warning]
                [Group: Undecoded]
    ApplicationId: 3GPP SGd (16777313)
    Hop-by-Hop Identifier: 0x00000005
    End-to-End Identifier: 0x4ec62f38
    [Answer In: 6]
    AVP: Session-Id(263) l=33 f=-M- val=smscHalys;1509618349;T4;0
        AVP Code: 263 Session-Id
        AVP Flags: 0x40
            0... .... = Vendor-Specific: Not set
            .1.. .... = Mandatory: Set
            ..0. .... = Protected: Not set
            ...0 .... = Reserved: Not set
            .... 0... = Reserved: Not set
            .... .0.. = Reserved: Not set
            .... ..0. = Reserved: Not set
            .... ...0 = Reserved: Not set
        AVP Length: 33
        Session-Id: smscHalys;1509618349;T4;0
        Padding: 000000
    AVP: Auth-Session-State(277) l=12 f=-M- val=NO_STATE_
    MAINTAINED (1)
        AVP Code: 277 Auth-Session-State
        AVP Flags: 0x40
            0... .... = Vendor-Specific: Not set
            .1.. .... = Mandatory: Set
            ..0. .... = Protected: Not set
            ...0 .... = Reserved: Not set
            .... 0... = Reserved: Not set
            .... .0.. = Reserved: Not set
            .... ..0. = Reserved: Not set
            .... ...0 = Reserved: Not set
        AVP Length: 12
        Auth-Session-State: NO_STATE_MAINTAINED (1)
    AVP: Origin-Host(264) l=47 f=-M- val=smsc1.epc.mnc094.mcc208.
    3gppnetwork.org
        AVP Code: 264 Origin-Host
        AVP Flags: 0x40
            0... .... = Vendor-Specific: Not set
            .1.. .... = Mandatory: Set
            ..0. .... = Protected: Not set
            ...0 .... = Reserved: Not set
            .... 0... = Reserved: Not set
```

```
            .... .0.. = Reserved: Not set
            .... ..0. = Reserved: Not set
            .... ...0 = Reserved: Not set
        AVP Length: 47
        Origin-Host: smsc1.epc.mnc094.mcc208.3gppnetwork.org
        Padding: 00
AVP: Origin-Realm(296) l=41 f=-M- val=epc.mnc094.mcc208.
3gppnetwork.org
        AVP Code: 296 Origin-Realm
        AVP Flags: 0x40
            0... .... = Vendor-Specific: Not set
            .1.. .... = Mandatory: Set
            ..0. .... = Protected: Not set
            ...0 .... = Reserved: Not set
            .... 0... = Reserved: Not set
            .... .0.. = Reserved: Not set
            .... ..0. = Reserved: Not set
            .... ...0 = Reserved: Not set
        AVP Length: 41
        Origin-Realm: epc.mnc094.mcc208.3gppnetwork.org
        Padding: 000000
AVP: Destination-Host(293) l=46 f=-M- val=mme1.epc.mnc094.mcc208.
3gppnetwork.org
        AVP Code: 293 Destination-Host
        AVP Flags: 0x40
            0... .... = Vendor-Specific: Not set
            .1.. .... = Mandatory: Set
            ..0. .... = Protected: Not set
            ...0 .... = Reserved: Not set
            .... 0... = Reserved: Not set
            .... .0.. = Reserved: Not set
            .... ..0. = Reserved: Not set
            .... ...0 = Reserved: Not set
        AVP Length: 46
        Destination-Host: mme1.epc.mnc094.mcc208.3gppnetwork.org
        Padding: 0000
AVP: Destination-Realm(283) l=41 f=-M- val=epc.mnc094.mcc208.
3gppnetwork.org
        AVP Code: 283 Destination-Realm
        AVP Flags: 0x40
            0... .... = Vendor-Specific: Not set
            .1.. .... = Mandatory: Set
            ..0. .... = Protected: Not set
            ...0 .... = Reserved: Not set
            .... 0... = Reserved: Not set
            .... .0.. = Reserved: Not set
            .... ..0. = Reserved: Not set
            .... ...0 = Reserved: Not set
        AVP Length: 41
        Destination-Realm: epc.mnc094.mcc208.3gppnetwork.org
        Padding: 000000
AVP: User-Name(1) l=23 f=-M- val=208940000001014
        AVP Code: 1 User-Name
```

```
            AVP Flags: 0x40
                0... .... = Vendor-Specific: Not set
                .1.. .... = Mandatory: Set
                ..0. .... = Protected: Not set
                ...0 .... = Reserved: Not set
                .... 0... = Reserved: Not set
                .... .0.. = Reserved: Not set
                .... ..0. = Reserved: Not set
                .... ...0 = Reserved: Not set
            AVP Length: 23
            User-Name: 208940000001014
            Padding: 00
        AVP: Unknown(3300) l=23 f=VM- vnd=TGPP val=3333393730363730303133
            AVP Code: 3300 Unknown
                Unknown AVP 3300 (vendor=3GPP), if you know what this is
                you can add it to dictionary.xml
                    [Expert Info (Warning/Undecoded): Unknown AVP
    3300 (vendor=3GPP), if you know what this is you can add it to
    dictionary.xml]
                        [Unknown AVP 3300 (vendor=3GPP), if you know what
                        this is you can add it to dictionary.xml]
                        [Severity level: Warning]
                        [Group: Undecoded]
            AVP Flags: 0xc0
                1... .... = Vendor-Specific: Set
                .1.. .... = Mandatory: Set
                ..0. .... = Protected: Not set
                ...0 .... = Reserved: Not set
                .... 0... = Reserved: Not set
                .... .0.. = Reserved: Not set
                .... ..0. = Reserved: Not set
                .... ...0 = Reserved: Not set
            AVP Length: 23
            AVP Vendor Id: 3GPP (10415)
            Value: 3333393730363730303133
            Padding: 00
        AVP: Unknown(3301) l=32 f=VM- vnd=TGPP val=040
        c913379600710f3480471112001529400 0100
            AVP Code: 3301 Unknown
                Unknown AVP 3301 (vendor=3GPP), if you know what this is
    you can add it to dictionary.xml

                    [Expert Info (Warning/Undecoded): Unknown AVP
    3301 (vendor=3GPP), if you know what this is you can add
    it to dictionary.xml]
                        [Unknown AVP 3301 (vendor=3GPP), if you know what
    this is you can add it to dictionary.xml]
                        [Severity level: Warning]
                        [Group: Undecoded]
            AVP Flags: 0xc0
                1... .... = Vendor-Specific: Set
                .1.. .... = Mandatory: Set
                ..0. .... = Protected: Not set
```

```
                ...0 .... = Reserved: Not set
                .... 0... = Reserved: Not set
                .... .0.. = Reserved: Not set
                .... ..0. = Reserved: Not set
                .... ...0 = Reserved: Not set
           AVP Length: 32
           AVP Vendor Id: 3GPP (10415)
           Value: 040c913379600710f3480471112001529400010
```

11.4.3.1 Uplink NAS transport request MME → UE and downlink NAS transport UE → MME

We have SMS-DELIVER to the UE and then SMS-DELIVER-REPORT returned by the UE after the SMS delivery.

```
No. Time              Source           Destination    Protocol Length IMSI
2    11:25:49,666946 192.168.2.113 192.168.2.117 GSM SMS 134
Signalling Link Selector
    DownlinkNASTransport, Downlink NAS transport(DTAP) (SMS) CP-DATA (
    RP) RP-DATA (Network to MS)

Frame 2: 134 bytes on wire (1072 bits), 134 bytes captured (1072 bits)
on interface 0
Ethernet II, Src: Vmware_f7:e8:3b (00:0c:29:f7:e8:3b), Dst:
Airspan_17:48:c0 (00:a0:0a:17:48:c0)
Internet Protocol Version 4, Src: 192.168.2.113, Dst: 192.168.2.117
Stream Control Transmission Protocol, Src Port: 36412 (36412),
Dst Port: 36412 (36412)
S1 Application Protocol
    S1AP-PDU: initiatingMessage (0)
        initiatingMessage
            procedureCode: id-downlinkNASTransport (11)
            criticality: ignore (1)
            value
                DownlinkNASTransport
                    protocolIEs: 3 items
                        Item 0: id-MME-UE-S1AP-ID
                            ProtocolIE-Field
                                id: id-MME-UE-S1AP-ID (0)
                                criticality: reject (0)
                                value
                                    MME-UE-S1AP-ID: 928194286
                        Item 1: id-eNB-UE-S1AP-ID
                            ProtocolIE-Field
                                id: id-eNB-UE-S1AP-ID (8)
                                criticality: reject (0)
                                value
                                    ENB-UE-S1AP-ID: 53
                        Item 2: id-NAS-PDU
                            ProtocolIE-Field
                                id: id-NAS-PDU (26)
```

```
                                criticality: reject (0)
                                value
                                NAS-PDU:
                27d73f3fe913076223690120010607913379600710f30014...
                                Non-Access-Stratum (NAS)PDU
                                    0010 .... = Security header type:
Integrity protected and ciphered (2)
                                    .... 0111 = Protocol
discriminator: EPS mobility management messages (0x7)
                                    Message authentication code:
                                    0xd73f3fe9
                                    Sequence number: 19
                                    0000 .... = Security header type:
Plain NAS message, not security protected (0)
                                    .... 0111 = Protocol
discriminator: EPS mobility management messages (0x7)
                                    NAS EPS Mobility Management
Message Type: Downlink NAS transport (0x62)
                                    NAS message container
                                    Length: 35
                                    NAS message container content:
690120010607913379600710f30014040c913379600710f3...
                                        GSM A-I/F DTAP - CP-DATA
                                        Protocol
                                        Discriminator:
                                        SMS messages (9)
                                            .... 1001 =
Protocol discriminator: SMS messages (0x9)
                                            0... .... =
TI flag: allocated by sender
                                            .110 .... = TIO: 6
                                        DTAP Short Message
Service Message Type: CP-DATA (0x01)
                                        CP-User Data
                                            Length: 32
                                            RPDU (not
                                            displayed)
                                        GSM A-I/F RP - RP-DATA
                                        (Network to MS)
                                            Message Type
                                            RP-DATA (Network
                                            to MS)
                                            RP-Message
                                            Reference
                                                RP-Message
                                                Reference:
                                                0x06 (6)
                                            RP-Originator
                                            Address -
                                            (33970670013)
                                                Length: 7
                                                1... .... =
Extension: No Extension
```

```
                                                         .001 .... =
Type of number: International Number (0x1)
                                                         .... 0001 =
Numbering plan identification: ISDN/Telephony Numbering
(ITU-T Rec. E.164 / ITU-T Rec. E.163) (0x1)
                                                         Called Party
            BCD Number: 33970670013
                                                         RP-Destination
                                                         Address
                                                             Length: 0
                                                         RP-User Data
                                                             Length: 20
                                                             TPDU (not
                                                             displayed)
                                                     GSM SMS TPDU
                                                     (GSM 03.40)
                                                     SMS-DELIVER
                                                         0... .... =
TP-RP: TP Reply Path parameter is not set in this SMS SUBMIT/DELIVER
                                                         .0.. .... =
TP-UDHI: The TP UD field contains only the short message
                                                         ..0. .... =
TP-SRI: A status report shall not be returned to the SME
                                                         .... 0... =
TP-LP: The message has not been forwarded and is not a spawned message
                                                         .... .1.. =
TP-MMS: No more messages are waiting for the MS in this SC
                                                         .... ..00 =
TP-MTI: SMS-DELIVER (0)
                                                         TP-Originating-
                                                         Address -
                                                         (33970670013)
                                                             Length: 12
                                                             address
                                                             digits
                                                             1... ....
= Extension: No extension
                                                             .001 ....
= Type of number: International (1)
                                                             .... 0001
= Numbering plan: ISDN/telephone (E.164/E.163) (1)
                                                             TP-OA
Digits: 33970670013
                                                         TP-PID: 72
                                                             01.. ....
= Defines formatting for subsequent bits: 0x1
                                                             ..00 1000
= Message type: Device Triggering Short Message (8)
                                                         TP-DCS: 4
                                                             00.. ....
= Coding Group Bits: General Data Coding indication (0)
                                                             ..0. ....
= Text: Not compressed
```

```
                                                            ...0 ....
= Message Class: Reserved, no message class
                                                            .... 01..
= Character Set: 8 bit data (0x1)
                                                            .... ..00
= Message Class: Class 0 (0x0)
                                                            TP-Service
                                                            -Centre-
                                                            Time-
                                                            Stamp
                                                                Year:
                                                                17
                                                                Month:
                                                                11
                                                                Day: 2
                                                                Hour:
                                                                10
Minutes: 25
Seconds: 49
Timezone: GMT + 0 hours 0 minutes
                                                            TP-User-
Data-Length: (1) depends on Data-Coding-Scheme
                                                            TP-User-
                                                            Data
                                                                SMS
                                                                body:
                                                                00
```

```
No. Time             Source        Destination   Protocol Length IMSI
3   11:25:49,821019 192.168.2.117 192.168.2.113 GSM SMS 130
Signalling Link Selector
    UplinkNASTransport, Uplink NAS transport(DTAP) (SMS) CP-DATA (RP)
    RP-ACK (MS to Network)

Frame 3: 130 bytes on wire (1040 bits), 130 bytes captured (1040 bits)
     on interface 0
Ethernet II, Src: Airspan_17:48:c0 (00:a0:0a:17:48:c0),
Dst: Vmware_f7:e8:3b (00:0c:29:f7:e8:3b)
Internet Protocol Version 4, Src: 192.168.2.117, Dst: 192.168.2.113
Stream Control Transmission Protocol, Src Port: 36412 (36412),
Dst Port: 36412 (36412)
S1 Application Protocol
    S1AP-PDU: initiatingMessage (0)
        initiatingMessage
            procedureCode: id-uplinkNASTransport (13)
            criticality: ignore (1)
            value
                UplinkNASTransport
                    protocolIEs: 5 items
                        Item 0: id-MME-UE-S1AP-ID
                            ProtocolIE-Field
                                id: id-MME-UE-S1AP-ID (0)
```

```
                                criticality: reject (0)
                                value
                                    MME-UE-S1AP-ID: 928194286
                        Item 1: id-eNB-UE-S1AP-ID
                            ProtocolIE-Field
                                id: id-eNB-UE-S1AP-ID (8)
                                criticality: reject (0)
                                value
                                    ENB-UE-S1AP-ID: 53
                        Item 2: id-NAS-PDU
                            ProtocolIE-Field
                                id: id-NAS-PDU (26)
                                criticality: reject (0)
                                value
                                    NAS-PDU:
                                    2759c6a4de1e076309e90106020641020000
                                        Non-Access-Stratum (NAS)PDU
                                        0010 .... = Security header
type: Integrity protected and ciphered (2)
                                        .... 0111 = Protocol
discriminator: EPS mobility management messages (0x7)
                                        Message authentication code:
0x59c6a4de
                                        Sequence number: 30
                                        0000 .... = Security header
type: Plain NAS message, not security protected (0)
                                        .... 0111 = Protocol
discriminator: EPS mobility management messages (0x7)
                                        NAS EPS Mobility Management
Message Type: Uplink NAS transport (0x63)
                                        NAS message container
                                            Length: 9
                                        NAS message container
content: e90106020641020000
                                            GSM A-I/F DTAP
                                            - CP-DATA
                                                Protocol
Discriminator: SMS messages (9)
                                                .... 1001 =
Protocol discriminator: SMS messages (0x9)
                                                1... .... =
TI flag: allocated by receiver
                                                .110 .... = TIO: 6
                                            DTAP Short Message
Service Message Type: CP-DATA (0x01)
                                                CP-User Data
                                                    Length: 6
                                                    RPDU (not
                                                    displayed)
                                            GSM A-I/F RP - RP-ACK
                                            (MS to Network)
                                                Message Type
                                                RP-ACK (MS
```

```
                                                  to Network)
                                                  RP-Message
                                                  Reference
                                                      RP-Message
                                                      Reference:
                                                      0x06 (6)
                                                  RP-User Data
                                                      Element ID:
                                                      0x41
                                                      Length: 2
                                                      TPDU (not
                                                      displayed)
                                          GSM SMS TPDU
                                          (GSM 03.40)
                                          SMS-DELIVER
                                          REPORT
                                              .0.. .... =
TP-UDHI: The TP UD field contains only the short message
                                              .... ..00 =
TP-MTI: SMS-DELIVER REPORT (0)
                                                  TP-Parameter-
Indicator: 0x00
                                                  0... .... =
Extension: No extension
                                                  .000 0... =
Reserved: 0
                                                  .... .0.. =
TP-UDL: Not Present
                                                  .... ..0. =
TP-DCS: Not Present
                                                  .... ...0 =
TP-PID: Not Present

                    Item 3: id-EUTRAN-CGI
                        ProtocolIE-Field
                            id: id-EUTRAN-CGI (100)
                            criticality: ignore (1)
                            value
                                EUTRAN-CGI
                                    pLMNidentity: 02f849
                                    Mobile Country Code (MCC):
                                    France (208)
                                    Mobile Network Code (MNC): Halys (94)
                                    cell-ID: 0000b010 [bit length 28,
4 LSB pad bits, 0000 0000 0000 0000 1011 0000 0001 .... decimal
value 2817]
                    Item 4: id-TAI
                        ProtocolIE-Field
                            id: id-TAI (67)
                            criticality: ignore (1)
                            value
                                TAI
                                    pLMNidentity: 02f849
```

```
                                Mobile Country Code (MCC):
                                France (208)
                                Mobile Network Code (MNC): Halys (94)
                                tAC: 1111 (0x0457)

No. Time Source Destination Protocol Length IMSI Signalling
Link Selector Info
    6 11:25:49,824159 192.168.2.114 192.168.2.116 DIAMETER 266
cmd=Unknown Answer(8388646) flags=-P-- appl=3GPP SGd(16777313) h2h=5
e2e=4ec62f38 |

Frame 6: 266 bytes on wire (2128 bits), 266 bytes captured (2128 bits)
on interface 0
Ethernet II, Src: Vmware_58:44:63 (00:0c:29:58:44:63), Dst:
Vmware_ea:40:0f (00:0c:29:ea:40:0f)
Internet Protocol Version 4, Src: 192.168.2.114, Dst: 192.168.2.116
Stream Control Transmission Protocol, Src Port: 3873 (3873),
Dst Port: 3873 (3873)
Diameter Protocol
    Version: 0x01
    Length: 204
    Flags: 0x40, Proxyable
        0... .... = Request: Not set
        .1.. .... = Proxyable: Set
        ..0. .... = Error: Not set
        ...0 .... = T(Potentially re-transmitted message): Not set
        .... 0... = Reserved: Not set
        .... .0.. = Reserved: Not set
        .... ..0. = Reserved: Not set
        .... ...0 = Reserved: Not set
    Command Code: 8388646 Unknown Transfer Forward Answer
        Unknown command, if you know what this is you can add it to
dictionary.xml
            [Expert Info (Warning/Undecoded): Unknown command, if you
know what this is you can add it to dictionary.xml]
                [Unknown command, if you know what this is you can
add it to dictionary.xml]
                [Severity level: Warning]
                [Group: Undecoded]
    ApplicationId: 3GPP SGd (16777313)
    Hop-by-Hop Identifier: 0x00000005
    End-to-End Identifier: 0x4ec62f38
    [Request In: 5]
    [Response Time: 0.583835000 seconds]
    AVP: Session-Id(263) l=33 f=-M- val=smscHalys;1509618349;T4;0
        AVP Code: 263 Session-Id
        AVP Flags: 0x40
            0... .... = Vendor-Specific: Not set
            .1.. .... = Mandatory: Set
            ..0. .... = Protected: Not set
            ...0 .... = Reserved: Not set
            .... 0... = Reserved: Not set
            .... .0.. = Reserved: Not set
```

```
                    .... ..0. = Reserved: Not set
                    .... ...0 = Reserved: Not set
            AVP Length: 33
            Session-Id: smscHalys;1509618349;T4;0
            Padding: 000000
    AVP: Vendor-Specific-Application-Id(260) l=32 f=-M-
            AVP Code: 260 Vendor-Specific-Application-Id
            AVP Flags: 0x40
                    0... .... = Vendor-Specific: Not set
                    .1.. .... = Mandatory: Set
                    ..0. .... = Protected: Not set
                    ...0 .... = Reserved: Not set
                    .... 0... = Reserved: Not set
                    .... .0.. = Reserved: Not set
                    .... ..0. = Reserved: Not set
                    .... ...0 = Reserved: Not set
            AVP Length: 32
            Vendor-Specific-Application-Id:
0000010a4000000c000028af000001024000000c01000061
                AVP: Vendor-Id(266) l=12 f=-M- val=10415
                    AVP Code: 266 Vendor-Id
                    AVP Flags: 0x40
                        0... .... = Vendor-Specific: Not set
                        .1.. .... = Mandatory: Set
                        ..0. .... = Protected: Not set
                        ...0 .... = Reserved: Not set
                        .... 0... = Reserved: Not set
                        .... .0.. = Reserved: Not set
                        .... ..0. = Reserved: Not set
                        .... ...0 = Reserved: Not set
                    AVP Length: 12
                    Vendor-Id: 10415
                    VendorId: 3GPP (10415)
                AVP: Auth-Application-Id(258) l=12 f=-M- val=3GPP
SGd (16777313)
                    AVP Code: 258 Auth-Application-Id
                    AVP Flags: 0x40
                        0... .... = Vendor-Specific: Not set
                        .1.. .... = Mandatory: Set
                        ..0. .... = Protected: Not set
                        ...0 .... = Reserved: Not set
                        .... 0... = Reserved: Not set
                        .... .0.. = Reserved: Not set
                        .... ..0. = Reserved: Not set
                        .... ...0 = Reserved: Not set
                    AVP Length: 12
                    Auth-Application-Id: 3GPP SGd (16777313)
    AVP: Result-Code(268) l=12 f=-M- val=DIAMETER_SUCCESS (2001)
            AVP Code: 268 Result-Code
            AVP Flags: 0x40
                    0... .... = Vendor-Specific: Not set
                    .1.. .... = Mandatory: Set
                    ..0. .... = Protected: Not set
```

```
                ...0 .... = Reserved: Not set
                .... 0... = Reserved: Not set
                .... .0.. = Reserved: Not set
                .... ..0. = Reserved: Not set
                .... ...0 = Reserved: Not set
        AVP Length: 12
        Result-Code: DIAMETER_SUCCESS (2001)
    AVP: Auth-Session-State(277) l=12 f=-M- val=NO_STATE_
MAINTAINED (1)
        AVP Code: 277 Auth-Session-State
        AVP Flags: 0x40
            0... .... = Vendor-Specific: Not set
            .1.. .... = Mandatory: Set
            ..0. .... = Protected: Not set
            ...0 .... = Reserved: Not set
            .... 0... = Reserved: Not set
            .... .0.. = Reserved: Not set
            .... ..0. = Reserved: Not set
            .... ...0 = Reserved: Not set
        AVP Length: 12
        Auth-Session-State: NO_STATE_MAINTAINED (1)
    AVP: Origin-Host(264) l=46 f=-M- val=mme1.epc.mnc094.mcc208.
3gppnetwork.org
        AVP Code: 264 Origin-Host
        AVP Flags: 0x40
            0... .... = Vendor-Specific: Not set
            .1.. .... = Mandatory: Set
            ..0. .... = Protected: Not set
            ...0 .... = Reserved: Not set
            .... 0... = Reserved: Not set
            .... .0.. = Reserved: Not set
            .... ..0. = Reserved: Not set
            .... ...0 = Reserved: Not set
        AVP Length: 46
        Origin-Host: mme1.epc.mnc094.mcc208.3gppnetwork.org
        Padding: 0000
    AVP: Origin-Realm(296) l=41 f=-M- val=epc.mnc094.mcc208.
3gppnetwork.org
        AVP Code: 296 Origin-Realm
        AVP Flags: 0x40
            0... .... = Vendor-Specific: Not set
            .1.. .... = Mandatory: Set
            ..0. .... = Protected: Not set
            ...0 .... = Reserved: Not set
            .... 0... = Reserved: Not set
            .... .0.. = Reserved: Not set
            .... ..0. = Reserved: Not set
            .... ...0 = Reserved: Not set
        AVP Length: 41
        Origin-Realm: epc.mnc094.mcc208.3gppnetwork.org
        Padding: 000000
```

11.4.4　Delivery report request (DRR) and answer (DRA) SMSC → MTC-IWF T4/diameter

```
No. Time            Source          Destination    Protocol Length IMSI
7   11:25:49,833884 192.168.2.116 192.168.2.114 DIAMETER 402 SACK
Signalling Link Selector Info
    cmd=3GPP-Delivery-Report Request(8388644) flags=R--- appl=3GPP T4
    (16777311) h2h=6 e2e=4ec62f39 |

Frame 7: 402 bytes on wire (3216 bits), 402 bytes captured (3216 bits)
on interface 0
Ethernet II, Src: Vmware_ea:40:0f (00:0c:29:ea:40:0f),
Dst: Vmware_58:44:63 (00:0c:29:58:44:63)
Internet Protocol Version 4, Src: 192.168.2.116,
Dst: 192.168.2.114
Stream Control Transmission Protocol, Src Port: 3873 (3873),
Dst Port: 3873 (3873)
Diameter Protocol
    Version: 0x01
    Length: 324
    Flags: 0x80, Request
        1... .... = Request: Set
        .0.. .... = Proxyable: Not set
        ..0. .... = Error: Not set
        ...0 .... = T(Potentially re-transmitted message): Not set
        .... 0... = Reserved: Not set
        .... .0.. = Reserved: Not set
        .... ..0. = Reserved: Not set
        .... ...0 = Reserved: Not set
    Command Code: 8388644 3GPP-Delivery-Report Request
    ApplicationId: 3GPP T4 (16777311)
    Hop-by-Hop Identifier: 0x00000006
    End-to-End Identifier: 0x4ec62f39
    AVP: Session-Id(263) l=33 f=-M- val=smscHalys;1509618349;T4;0
        AVP Code: 263 Session-Id
        AVP Flags: 0x40
            0... .... = Vendor-Specific: Not set
            .1.. .... = Mandatory: Set
            ..0. .... = Protected: Not set
            ...0 .... = Reserved: Not set
            .... 0... = Reserved: Not set
            .... .0.. = Reserved: Not set
            .... ..0. = Reserved: Not set
            .... ...0 = Reserved: Not set
        AVP Length: 33
        Session-Id: smscHalys;1509618349;T4;0
        Padding: 000000
    AVP: Auth-Session-State(277) l=12 f=-M- val=NO_STATE_
    MAINTAINED (1)
        AVP Code: 277 Auth-Session-State
        AVP Flags: 0x40
            0... .... = Vendor-Specific: Not set
```

```
            .1.. .... = Mandatory: Set
            ..0. .... = Protected: Not set
            ...0 .... = Reserved: Not set
            .... 0... = Reserved: Not set
            .... .0.. = Reserved: Not set
            .... ..0. = Reserved: Not set
            ... ...0 = Reserved: Not set
        AVP Length: 12
        Auth-Session-State: NO_STATE_MAINTAINED (1)
    AVP: Origin-Host(264) l=47 f=-M- val=smsc1.epc.mnc094.mcc208.
    3gppnetwork.org
        AVP Code: 264 Origin-Host
        AVP Flags: 0x40
            0... .... = Vendor-Specific: Not set
            .1.. .... = Mandatory: Set
            ..0. .... = Protected: Not set
            ...0 .... = Reserved: Not set
            .... 0... = Reserved: Not set
            .... .0.. = Reserved: Not set
            .... ..0. = Reserved: Not set
            .... ...0 = Reserved: Not set
        AVP Length: 47
        Origin-Host: smsc1.epc.mnc094.mcc208.3gppnetwork.org
        Padding: 00
    AVP: Origin-Realm(296) l=41 f=-M- val=epc.mnc094.mcc208.
    3gppnetwork.org
        AVP Code: 296 Origin-Realm
        AVP Flags: 0x40
            0... .... = Vendor-Specific: Not set
            .1.. .... = Mandatory: Set
            ..0. .... = Protected: Not set
            ...0 .... = Reserved: Not set
            .... 0... = Reserved: Not set
            .... .0.. = Reserved: Not set
            .... ..0. = Reserved: Not set
            .... ...0 = Reserved: Not set
        AVP Length: 41
        Origin-Realm: epc.mnc094.mcc208.3gppnetwork.org
        Padding: 000000
    AVP: Destination-Host(293) l=46 f=-M- val=mtc1.epc.mnc094.mcc208.
    3gppnetwork.org
        AVP Code: 293 Destination-Host
        AVP Flags: 0x40
            0... .... = Vendor-Specific: Not set
            .1.. .... = Mandatory: Set
            ..0. .... = Protected: Not set
            ...0 .... = Reserved: Not set
            .... 0... = Reserved: Not set
            .... .0.. = Reserved: Not set
            .... ..0. = Reserved: Not set
            .... ...0 = Reserved: Not set
        AVP Length: 46
        Destination-Host: mtc1.epc.mnc094.mcc208.3gppnetwork.org
```

```
        Padding: 0000
   AVP: Destination-Realm(283) l=41 f=-M- val=epc.mnc094.mcc208.
   3gppnetwork.org
        AVP Code: 283 Destination-Realm
        AVP Flags: 0x40
             0... .... = Vendor-Specific: Not set
             .1.. .... = Mandatory: Set
             ..0. .... = Protected: Not set
             ...0 .... = Reserved: Not set
             .... 0... = Reserved: Not set
             .... .0.. = Reserved: Not set
             .... ..0. = Reserved: Not set
             .... ...0 = Reserved: Not set
        AVP Length: 41
        Destination-Realm: epc.mnc094.mcc208.3gppnetwork.org
        Padding: 000000
   AVP: User-Identifier(3102) l=36 f=VM- vnd=TGPP
        AVP Code: 3102 User-Identifier
        AVP Flags: 0xc0
             1... .... = Vendor-Specific: Set
             .1.. .... = Mandatory: Set
             ..0. .... = Protected: Not set
             ...0 .... = Reserved: Not set
             .... 0... = Reserved: Not set
             .... .0.. = Reserved: Not set
             .... ..0. = Reserved: Not set
             .... ...0 = Reserved: Not set
        AVP Length: 36
        AVP Vendor Id: 3GPP (10415)
        User-Identifier:
   00000001400000173230383934303030303030313031340 0
           AVP: User-Name(1) l=23 f=-M- val=208940000001014
               AVP Code: 1 User-Name
               AVP Flags: 0x40
                    0... .... = Vendor-Specific: Not set
                    .1.. .... = Mandatory: Set
                    ..0. .... = Protected: Not set
                    ...0 .... = Reserved: Not set
                    .... 0... = Reserved: Not set
                    .... .0.. = Reserved: Not set
                    .... ..0. = Reserved: Not set
                    .... ...0 = Reserved: Not set
               AVP Length: 23
               User-Name: 208940000001014
               Padding: 00
   AVP: SM-RP-SMEA(3309) l=18 f=VM- vnd=TGPP val=3379600710f3
        AVP Code: 3309 SM-RP-SMEA
        AVP Flags: 0xc0
             1... .... = Vendor-Specific: Set
             .1.. .... = Mandatory: Set              ..0. .... =
Protected: Not set
             ...0 .... = Reserved: Not set
             .... 0... = Reserved: Not set
```

```
          .... .0.. = Reserved: Not set
          .... ..0. = Reserved: Not set
          .... ...0 = Reserved: Not set
     AVP Length: 18
     AVP Vendor Id: 3GPP (10415)
     SM-RP-SMEA: 3379600710f3
     Padding: 0000
   AVP: SM-Delivery-Outcome-T4(3200) l=16 f=VM- vnd=TGPP
val=SUCCESSFUL_TRANSFER (2)
     AVP Code: 3200 SM-Delivery-Outcome-T4
     AVP Flags: 0xc0
          1... .... = Vendor-Specific: Set
          .1.. .... = Mandatory: Set
          ..0. .... = Protected: Not set
          ...0 .... = Reserved: Not set
          .... 0... = Reserved: Not set
          .... .0.. = Reserved: Not set
          .... ..0. = Reserved: Not set
          .... ...0 = Reserved: Not set
     AVP Length: 16
     AVP Vendor Id: 3GPP (10415)
     SM-Delivery-Outcome-T4:           SUCCESSFUL_TRANSFER (2) ..
```

In this test as there is no MTC-IWF you do not see the DRA answer

11.5 Traces of a recursive DNS request

The DNS architecture and protocol (UDP port 53) are the key of the addressing scheme of the Internet which explains the ease to create quickly world accessible services. It is also used in the GRX private Internet of the mobile operators. The addressing is hierarchical in the reverse order of the domain names such as:

internet.mnc001.mcc222.gprs or

tenders.renault.fr (root DNS 2, then .fr DNS 4, then Renault DNS 6, then tenders website 11).

All the requests start by interrogating the "root DNS" or a national copy under the control and management of ICAN which has delegated the running to Neustar, a private company. ICAN has thus a huge control over the running of all the Internet services as it is possible by suppressing an entry in the root DNS to cut the access to any service.

The various national DNS such as .fr can also block the access to any of the services which as a .fr domain name termination.

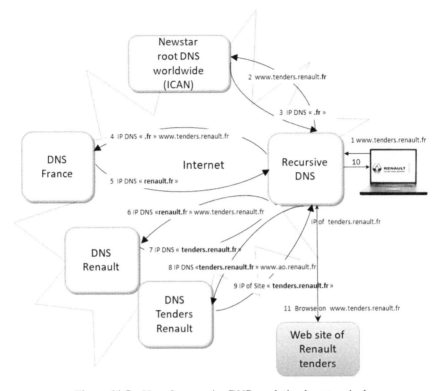

Figure 11.5   Use of a recursive DNS resolution by a terminal.

The browser on the right can be:

- a PC, the IP of the Recursive DNS is provided by the DHCP request to the access provider;
- a mobile handset, the IP of the Recursive DNS is provided by the Create Session Response when the handset attaches to the 4G network.

Figure 11.5 illustrates the use of DNS by the terminal when the data session is established.

DNS requests are also used by SGSN or MME, by SGW (or GTP hub) to find the IP of a GGSN-PGW serving a particular APN. The DNS address for mobiles is assigned by the GGSN-PGW in the Create PDP Context Response (GTPv1 3G) or Create Session Response (GTPv2 4G) as well as the IP of the terminals when it is not static.

```
(128):End User Address
                 L(TLV type) = 6
                   PDP Type Organisation 1
                   PDP Type Number: (33):IPv4
                 Data: 199.255.7.18 // IP assigned by the GGSN-PGW

              (132):Protocol Configuration Options // DNS assigned
              by the GGSN-PGW
                L(TLV type) = 28
                Data (V1):
                Extension :(1) Configuration protocol TS 24.008
                rel 11 page 555: 0x80
                Configuration Protocol : PPP for use with IP PDP
                type or IP PDN type (0)
                Configuration Protocol options list
                (C023) Protocol or Container ID : PPP Password
                Authentication Protocol(PAP) RFC 1334
                Length = 5
                               Code : (2):Authentication-Ack
                               Identifier : 0
                               Length : 5
                                Ack or Nak Message Length : 0
                               Data :
                (8021) Protocol or Container ID : PPP IP Control
                Protocol(PPP) RFC 1661
                Length = 16
                               Code : (3):Config-Nak
                               Identifier : 0
                               Length : 16

                               Primary DNS
                                 Length 6
                                 IP:
                192.168.2.114 // DNS of the MNO
                               Secondary DNS
                                 Length 6
                                 IP: 8.8.8.8
```

12

Conclusion: Full LTE for Security Forces, When?

It is not worthy of the educated man to waste his time in slave's work, that is calculus, with can be performed by the first one to come which has a calculator.

Wilhem Leibnitz (1646–1716), invented in 1673 a mechanical arithmetic calculator able of the four operations.

Comment: laborious teaching of integration or derivation rules is not worth, most 2018 formal calculus software do it perfectly.

The LTE-based security networks' perspective must be assessed by comparing to the currently used (2018) technologies, mostly the Tetra family which correctly handles the direct call function purely talkie-walkie considered as essential. Hence even in the not so frequent (2018) case where LTE is deployed, with great satisfaction for the data services, the security forces keep a second Tetra terminal.

Definitely the complete replacement by LTE waits the venue of direct call capable terminals, hence of chipset with a dual mode: terminal and station (to receive a direct call). This is a small market for manufacturers which volumes are hundreds of millions per year. The 2021 limit for certain Tetra contracts will certainly need to be extended for pure security forces (intervention police or civil security forces) and army forces.

What is the impact of 5G [12.5, 12.6] in this evolution toward a single LTE terminal for security forces? It is difficult to see any. The real 4G bandwidth (even if throttled by a lack of PMR allocated resources) is already quite adequate for the most demanding applications such as push-to-talk visio, provided that multicasting is used. It is always cheaper to optimize the use of resources by an evolution of the infrastructure than adding to them. 5G is something in all cases welcomed by the current or new network vendors as 4G was; it needs a newly designed EPC, Figure 12.1 gives a rough

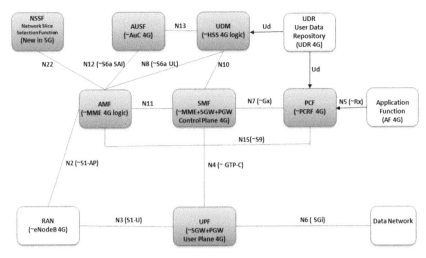

Figure 12.1   5G core network and equivalence with the known 4G network.

equivalence, and this is not a huge task for those which master all the 4G bricks.

The most important question is the questionable social utility of a very large bandwidth. When discussing "Fiber To The Home" instead of a radio access network, the conclusion is the same. A large bandwidth favors a waste of time by the TV customers (amusements and shows, instead of education). The current unicasting of channels is a waste from a common sense point of view: 10 persons watching a game at the same time use 10 channels, multicasting only one. The reduction of the number of channels, through the use of multicasting, would suppress the need to invest in FTTH.

In most of the world a major factor to be elected to a major public office requires not a high education or brilliant shared political conviction but the medias' support which have become over powerful. Many were not the best pupils in class; this is why one hears so much of artificial intelligence, a way of reducing the admiration for real intelligence and for the painful learning. The root of the speech recognition now successful techniques were invented (Markov chain learning) in the 80s. Deep leaning just used many times more powerful computers and larger leaning databases, where is the intelligence? The great Gauss took two years to learn to read, write and talk Russian aged 68, he will look as a foolish elder to peoples which shortly will have a practical multi-language speaking interpreter in their smartphone. No one except that a few alphas will need to learn foreign languages, history, geography, science, literature or even elementary divisions (this was not what Leibnitz aimed at in the quote). Practical IA and solidarity among them will allow the gammas and below categories to avoid major learning effort while being totally in the hands of the alphas and betas; this story was told already, washed by world culture no one will be able to rise for a cause as did the author of "1984".

Arnaud Henry-Labordère
July 2018.

References

[12.1] George Orwell, Nineteen Eighty-Four, Secker and Warburg, 1949, *his other most well know book is "Animals' farm"*.

[12.2] George Orwell, Homage to Catalonia, Secker and Warburg, 1938, *his enrolment in the international brigades and the internal fight between the organised communists, the POUM socialists and the CFT anarchists.*

[12.3] George Orwell, The road to Wigan Pier, Penguin Books, 1962. *The workers' misery in the 1930s.*

[12.4] Gauss, Génies Mathématiques, RBA Coleccionables, fév 2018, page 150, *a recent series of more than 20 volumes originally published in spain with great names, many successful foreign translations.*

[12.5] 3GPP TS 23.501 v15.1.0 (2018–03), "3rd Generation Partnership Project; Technical Specification Group Services and System Aspects; System Architecture for the 5G System; Stage 2," Release 15, *the 5G core network architecture is new compared to 4G. Start with that to design a 5G EPC.*

[12.6] 3GPP TS 23.502 v15.1.0 (2018–03), "3rd Generation Partnership Project; Technical Specification Group Services and System Aspects; Procedures for the 5G System; Stage 2," Release 15.

[12.7] 3GPP TS 23.503 v15.1.0 (2018–03), "3rd Generation Partnership Project; Technical Specification Group Services and System Aspects; Policy and Charging Control Framework for the 5G System; Stage 2," Release 15. *5G is service oriented, with sophisticated charging schemes.*

Abbreviations and Acronyms

A	
A-GNSS	Assisted Global Navigation Satellite System
A-GPS	Assisted GPS
A3	Authentication algorithm A3 used in 2G GSM with a challenge between the HLR and the SIM card using a random number and a result computed with a shared secret Ki between the HLR (AuC) and the SIM card.
A38	A single algorithm performing the functions of A3 and A8
A5/1	Encryption algorithm A5/1
A5/2	Encryption algorithm A5/2
A5/X	Encryption algorithm A5/0-7
A8	Ciphering key generating algorithm A8
AA19	Standard GSM contract between 2 operators for the charging of the SMS-MT sent to their own subscribers by the other.
AALBsec	Active-Active Load Balancing Secured
AB	Access Burst
ABP	Activation-by-Personalization, *one of the LoRa modes*
AC	Access Class (C0 to C15)
	Application Context
AC	Autorité de Certification (CA), *which issued the certificates*
ACC	Automatic Congestion Control
ACCH	Associated Control Channel
ACELP	Algebraic Code Excited Linear Predictive, *codec assez ancien mais performant pour le faible débit du codec audio TETRA 4,737 kb/sec*
ACK	ACKnowledgment
ACM	Accumulated Call Meter (a zone of a SIM card)
	Address Complete Message (Response to a ISUP Call setup)
ACMmax	Maximum of the Accumulated Call Meter
ACSE	Association Control Service Element
ACU	Antenna Combining Unit
ADC	ADministration Center
ADC	Application Detection and Control, Fonction DPI dans un PCEF
ADC	Active Directory Connector
	Analog to Digital Converter
ADD	Automatic Device Detection (inclusion of IMEIsv in Update Location)
ADN	Abbreviated Dialling Number

293

ADPCM	Adaptive Differential Pulse Code Modulation
AE	Application Entity
AEC	Acoustic Echo Control
AEF	Additional Elementary Functions
AES	Advanced Encryption Standard, *algorithme symétrique dû aux belges Rijmen et Daemen d'où le nom Rijndael donné aussi. Utilise des clés de 128 bits et est notamment utilisé dans l'algorithme milenage d'authentification des mobiles par le HLR-HSS, Ki étant la clé d'authentification propre à chaque abonné. Remplace l'ancien algorithme DES à seulement 56 bits*
AF	Application Function. En IMS, terme générique qui comprend le P-CSCF et l'AS (Application Server), par exemple TAS communiquant en SIP. Le P-CSCF est relié en Rx avec le PCRF et Cx avec le HLR-HSS.
AGCH	Access Grant Channel
AKA	Authentication and Key Agreement
AKAv1-MD5	*see chapter 2. The authentication scheme used in IMS between the terminal and the HSS, the P-CSCF plays the role of a VLR 3G comparing the two XRES computed by the HSS and the UE. This is not the same scheme as the one used in ordinary SIP OTT.* Ai Action indicator
AMF	Authentication Management Field, *16 bits dans l'AUTN du vecteur d'authentication. Le bit 0 ("separation bit" doît être à 1 pour le LTE sinon la SIM le refuse et l'eNodeB retourne EMM cause 20 (MAC failure)*
AMF	Access and Mobility Management Function, *rough 5G equivalent of the logic of an MME*
AMPS	Advanced Mobile Phone System(Analog mobile Radio system)
ANR	Automatic Neighbour Relation. *Instead of configuring each eNodeB with its neighbour list for the X2 Handover X2, the MME is interrogated by the eNodeB and provides this list*
ANDSF	Access Network Discovery and Selection Function, système de gestion des Access Class des différentes stations radio d'un réseau
ANSI	American National Standards Institute
ANSSI	Agence Nationale de Sécurité des Système d'Information (France)
AoC	Advice of Charge
AoCC	Advice of Charge Charging supplementary service
AoCI	Advice of Charge Information supplementary service
APLMN	Associated Public Land Mobile Network
APN	Access Provider Name
ARD	Access Restriction Data. *In the HLR-HSS subscriber profile, list of restricted technologies (GERAN 2G, EUTRAN 4G, etc.)*
ARFCN	Absolute Radio Frequency Channel Number, *un code correspondant à la fréquence utilisée par l'UE*
ARP	Alternative Roaming Providers, for the data roaming. *Notion introduced in the new European regulation for the Local Break-Out*

ARP	Allocation and Retetention Priority, Paramétre de Qos du HLR-HSS ou d'un GGSN-PGW fixant la priorité et la capacité à préempter ou à être pré-empté.
ASE	Application Service Element
AS	Autonomous System: a subset of the IP network with a common routing policy.
ASN	Autonomous System Number, a unique 16 bits number(IPV4) or 32 bits number(IPV6) defining the set of IP addresses of an operator for the purpose of configuring his "Border Gateway" and included in his IR21 document. The ASN is delivered at the same time as some public IP addresses are allocated to companies.
ASN.1	Abstract Syntax Notation One
ARFCN	Absolute Radio Frequency Channel Number
ARQ	Automatic ReQuest for retransmission
AS	Application Server (TAS=Telephony Application Server, par exemple IMS VoIP)
ASP	Application Service Provider (Content Provider for Internet services)
ATT	(flag) ATTach
AU	Access Unit
AuC	Authentication Center, part of an HLR-HSS which holds the security keys Ki and executes the COMP8 and Milenage authentication algorithm
AUSF	Authentication Server Function, 5 G equivalent of the AuC.
AUT(H)	AUThentication
AVP	Attribute Value Pair, name of the parameters in the RADIUS and DIAMETER protocols
B	
BA	BCCH Allocation
BAIC	Barring of All Incoming Calls supplementary service
BAOC	Barring of All Outgoing Calls supplementary service
BAP	Bon à Produire (english: Production agreement), *this is the administrative ordering process, with a cost, to the SIM manufacturer for a new release of a SIM cards with additional directories or files (not just a content change)*
BBERF	Bearer Binding and Event Reporting Function, function in the SGW responsible for bearer binding.
BCC	Base Transceiver Station (BTS) Color Code
BCCH	Broadcast Control CHannel
BCD	Binary Coded Decimal
BCF	Base station Control Function
BCIE	Bearer Capability Information Element
BD	Billing Domain, Charging with an OCS
BDU	Bien à Double Usage
BER	Bit Error Rate
BER	Basic Encoding Rule, Variante de codage ASN1 pour MAP, Camel

BEREC   Body of European Registrators for Electronic Communications
BFI   Bad Frame Indication
BGP   Border Gateway Protocol
BGW   Border Gateway. *Connects a PLMN to GRX*
BI   all Barring of Incoming call supplementary services
BIB   Backward Indicator Bit
BIC   Roam Barring of Incoming Calls when Roaming outside the home
               PLMN country supplementary service
BIP   Bearer-Independent Protocol, *Protocole pour SIM OTA utilisant la
               connexion GPRS, beaucoup plus rapide que les SMS, mais pas encore
               disponible dans tous les terminaux.*
Bm   Full-rate traffic channel
BM-SC   Broadcast and Multicast Service Center *(server used for the MBMS
               service, interfacing by the SGmb interface with the MBMS Gateway),
               quelques fabricants: Enensys, Expway.*
BN   Bit Number
BO   all Barring of Outgoing call supplementary, service
BOIC-exHC   Barring of Outgoing International Calls except those directed to the
               Home PLMN Country supplementary service
BPM   Broadcast Provisioning Manager, le système de gestion du BM-SC à
               ouvrir les canaux de broadcast pour un contenu donné ou une Service
               Area donnée
BPN   Bloc Primaire Numérique (Primary Numeric Block), French acronym
               for a 2Mb/sec E1
BS-   Basic Service (group)
               Bearer Service
BSG   Basic Service Group
BSC   Base Station Controller GSM 2
BSF   Bootstrapping Server Function, serveur indépendant des autres
               équipements pour leur identification sécurisée mutuelle. Intervient
               dans l'architecture LOCSIP
BSIC   Base transceiver Station Identity Code
BSIC   NCELL BSIC of an adjacent cell
BSN   Backward Sequence Number
BSS   Base Station System (GSM 2G)
BSSAP   Base Station System Application Part
BSSAP-LE   BSSAP with Location Extension
BSSMAP   Base Station System Management Application Part
BSSOMAP   Base Station System Operation and Maintenance Application Part
BTS   Base Transceiver Station GSM 2G

C
C   Conditional
C-PDS   CDMA Packet Data Service
C-TEID Co   commmon TEID, *pour le Multicast MMBS*

C-SGN CioT	Serving Gateway Node, combines MME, *SGW and PGW with the interfaces to optimize*
CA	Cell Allocation
CA	Autorité de certification *qui délivre les certificats*
CAI	Charge Advice Information
CAMEL	Customized Application for Mobile Network
CAMELize	Action to provision a CAMEL record in a subscriber's profile for other purpose than pre-payment in order to trigger an IN service, such as automatic APN correction.
CAN	Connectivity Access Network
CAT	Card application Toolkit, *contains the list of proactive comands*
CAT_TP	Card Application Toolkit Transport Protocol, *end-to-end between the SIM and the OTA server, according to ETSI TS 102 127.*
CB	Cell Broadcast
CBC	Cell Broadcast Center
	Ciphering Block Chaining (used in the 3DES algorithm)
CBCH	Cell Broadcast CHannel
CBMI	Cell Broadcast Message Identifier
CC	Country Code
	Content of Communication, *in Lawful Interception the "en clair" Content of Communication passed to the LEA's LEMF with the HI3 interface*
CC-API	Control Center API, en Tetrapol interface pour les SDS (SMS-MO et SMS-MT)
CCA	Credit Control Acknowledgment (Diameter Credit Control Application)
CCBS	Completion of Calls to Busy Subscriber supplementary service
CCCH	Common Control CHannel
CCF	Conditional Call Forwarding
CCH	Control CHannel
CCITT	Comité Consultatif International Télégraphique et Téléphonique (The International Telegraph and Telephone Consultative Committee)
CCM	Current Call Meter
CCP	Capability/Configuration Parameter
CCPE	Control Channel Protocol Entity
CCR	Credit Control Request (Diameter Credit Control Application)
CCA	Critical Communication Application, The Group Call Application in TETRA
CCS	Critical Communication System,
Cct	Circuit
CDF	Charging Data Function
CDMA	Code Division Multiple Access
CDN	Call Directory Number (the original MSISDN for a call to a ported-out number)
CDR	Call Detailed Record (billing record)
CDUR	Chargeable DURation

CED	called station identifier
CEIR	Central Equipment Identity Register
CEND	end of charge point
CEPT	Conférence des administrations Européennes des Postes et Télécommunications
CF	Conversion Facility
	all Call Forwarding services
CFB	Circuit Fall Back
cfb	Call Forwarding on mobile subscriber Busy supplementary service
cfnrc	Call Forwarding on mobile subscriber Not Reachable supplementary service
cfnry	Call Forwarding on No Reply supplementary service
cfu	Call Forwarding Unconditional supplementary service
CGI	Cell Global Identifier, *2G ou 3G MCC.MNC.LAC(16 bits),CI(16 bits)*
CHP	CHarging Point
CHV	Card Holder Verification information
CI	Cell Identity
	CUG Index
CIC	Circuit Identification Code
CIot	Cellular Internet of things
CIR	Carrier to Interference Ratio
CIS	Critically Important Services(administrations, utilities)
CKSN	Ciphering Key Sequence Number
CLI	Calling Line Identity
	Client-Identity, For LBS, E164 MSISDN of the Client Identitities authorized to perform a PSL
CLIP	Calling Line Identification Presentation supplementary service
CLIR	Calling Line Identification Restriction supplementary service
CLOUD ACT	Clarifying Lawful Overseas Use of Data(!). US Law of the 23 March 2018, rushed voted to counter the later coming EU GDPR law which was to become applicable in all UE the 25 May 2018. It makes an obligation to all US firms to transfer all user data (including European) to a US site, this is exactly opposite to the EU GDPR which forbids it.
CLS	Central Location Server, *Serveur de géolocalization TETRA*
CM	Connection Management
CMD	CoMmanD
CMM	Channel Mode Modify
CNF	CoNFirmation (Answer to a REQ(Request))
CN	Core Network
CNG	CalliNG tone
CNRC	Controlling RNC
CNTR	Counter (control for OTA SIM security)
COLI	COnnected Line Identity
COLP	COnnected Line identification Presentation supplementary service
COLR	COnnected Line identification Restriction supplementary service
COM	COMplete

COMP 128	Authentication and Ciphering algorithm used for A3 and A8 (GSM)
COMP 128-2	Improved algorithm used for UMTS
CONNACK	CONNect ACKnowledgment
C/R	Command/Response field bit
CRC	Cyclic Redundancy Check (3 bit)
CRE	Call RE establishment procedure
CRS	Cell Specific Reference Signal
CRX	CDMA Roaming ExChange (equivalent of GRX for CDMA data roaming)
CS	Domain Circuit Service Domain (includes MSCs)
CSCF	Call Session Control Function, Etablissement des sessions IMS.
CSFB	Circuit Switched Fall Back
CSG	Closed Subscriber Group
CSS	Closed Subscriber Group Server, can be part of the HLR-HSS giving a list of Cells authorized for a CSG
CSPDN	Circuit Switched Public Data Network
CSPN	Certificat de Sécurité Premier Niveau, Procédure de certifcation par l'ANSSI qu'un produit a subi des tests sécurité
CT	Call Transfer supplementary service
	Channel Tester
	Channel Type
CTF	Charging Trigger Function, MSC, SGSN, etc. which triggers the OCS
CTR	Common Technical Regulation
CUG	Closed User Group supplementary service
CW	Call Waiting supplementary service
D	
DA	Destination Address
DAC	Digital to Analog Converter
DAMPS	Digital AMPS (the TDMA mobile adio system)
DAO	Decentralized Autonomous Organization, la chaîne de blocs (*blockchain*) est une technologie nouvelle de stockage et de transmission d'informations, transparente, sécurisée, et fonctionnant sans organe central de contrôle. Utilisé dans le système de "bitcoin".
DAP	Data Authentication Pattern, *used for SmartCard loading*
DB	Dummy Burst
DCCH	Dedicated Control CHannel
DCE	Data Circuit terminating Equipment
DCF	Data Communication Function
DCN	Data Communication Network
DCS1800	Digital Cellular System at 1800MHz
DES	Data Encryption Standard (used for security of OTA SIM)
3DES	Triple DES (with 2 keys)
DET	DETach
DF	Delivery Functions
DGA	Direction Générale de l'Armement, Ministère de la Défense, France

DGSE	Direction Générale de la Sécurité Extérieure, Ministère de la Défense, France
DGSI	Direction Générale de la Sécurité Intérieure, Ministère de l'Intérieur, France
DHCP	Dynamic Host Configuration Protocol. The DHCP server assigns an IP address to a terminal through this protocol.
DIAMETER	an Authentication, Authorization and Charging Protocol which is more comprehensive than RADIUS
DISC	DISConnect
DL	Data Link (layer)
DL	DownLink
DLCI	Data Link Connection Identifier
DLD	Data Link Discriminator
Dm	Control channel (ISDN terminology applied to mobile service)
DMO	Direct Mode Operation, Appel direct entre 2 téléphones indépendamment du réseau d'accès
DMR	Digital Mobile Radio
DNIC	Data network identifier
DNS	Domain Name Server (accessed with the DNS protocol)
DNS	NAPTR, DNS Request type including the type of record requested for example the EPC domain. For diameter, the client and server DNS must support NAPTR.
DP	Dial/Dialled Pulse
	Destination Point (of an IN service)
DPC	Destination Point Code
DPSD	Direction de la Protection et de la Sécurité de la Défense, Ministère de la Défense, France
DRM	Digital Right Management
DRX	Discontinuous reception (mechanism)
DSCP	Differentiated Services Code Point. *Champ, de 6 bits utile dans les en-tête IP spécifiant le type de qualité de transport requise, RFC 2474*
DSE	Data Switching Exchange
DSI	Digital Speech Interpolation
DSS1	Digital Subscriber Signaling No1
DTAP	Direct Transfer Application Part
DTE	Data Terminal Equipment
DTLS	Datagram Transport Layer Security, for the MB2-U data transport
DTMF	Dual Tone Multi Frequency (signaling)
DTX	Discontinuous transmission (mechanism)
DVB-H	Digital Video Broadcast-Handheld (MULTICAST method for mobile TV)
E	
EA	External Alarms
EAB	Extended Access Barring, SIM card configuration decided by the HPLMN to control the network access.

EAL	Evaluation Assurance Level for secured systems approved by more than 20 countries.
EAL1	A low but cheap level of security, system functionnally tested
EAL4	Designed, tested and checked (highest level for most commercial products)
EAL4+	same as EAL4 with additional verifications of EAL5
EAL5	A high level of certification for secured communication systems semiformally designed and tested.
EAL5+	Can resist all physical attack of the electronic
EAL7	Highest level for government, formally verified designed and tested (no commercial product has (2015) EAL6 or EAL7.
EARFCN	Absolute Radio Frequency Channel Number for LTE carrier channel numbers
EAP	Extensible Authentication Protocol
EAP-AKA	EAP for 3rd Generation USIM cards
EAP-SIM	EAP for SIM cards
EBSG	Elementary Basic Service Group
ECID	Enhanced cell ID (using Timing Advance, Power Measurements, UE Rx-Tx Time Differences, etc.
ECM	Error Correction Mode (facsimile)
Ec/No	Ratio of energy per modulating bit to the noise spectral density
ECGI E-UTRAN	Cell Global Identifier, *4G cell MCC.MNC.ECI (sur 24 bits)*
ECI E-UTRAN	Cell Identifier, *24 bits.*
ECT	Explicit Call Transfer supplementary service
EDGE	Enhanced Data Rates for GSM Evolution ("2.75 intermediate generation allowing much higher rates without UMTS: 170 kbits/sec for visio)
EEL	Electric Echo Loss
EIA	Electronic Industries Equipment
EIR	Equipment Identity Register
EL	Echo Loss
EMC	ElectroMagnetic Compatibility
eMLPP	enhanced Multi-Level Precedence and Pre-emption service
EMMI	Electrical Man Machine Interface
ENUM	E164 Number Mapping
ePDG	evolved Packet Data Gateway, Gateway with an AAA server to secure a non-LTE access (e.g., WiFi) to the LTE PGW with the S2b interface.
EPROM	Erasable Programmable Read Only Memory
EPS	Evolved Packet system
ERP	Ear Reference Point
	Equivalent Radiated Power
ERR	ERRor
ESME	External Short Message Entity (an ASP or ISP) connected by SMPP
ESN	Electronic Serial Number
ESP	Encapsulating Security Payload, *pour le transport du MB2-U*
ETR	ETSI Technical Report

ETS	European Telecommunication Standard
ETSI	European Telecommunications Standards Institute
eUICC	Cartes SIM dernière génération *re-programmables*
E164	Format of the "ordinary" telephone numbers with a "Country Code" (CC) and a Network Destination Code(NDC)
E212	Format of the "IMSI" telephone numbers with a "Mobile Country Code" (MCC) and a Mobile Network Code (MNC)
E214	Format of a Destination Address, a mix of E164 and E212
E-OTD	Enhanced Observed Time Difference
F	
FA	Full Allocation
	Fax Adaptor
FAC	Final Assembly Code
FACCH	Fast Associated Control CHannel
FACCH/F	Fast Associated Control Channel/Full rate
FACCH/H	Fast Associated Control Channel/Half rate
FB	Frequency correction Burst
FCCH	Frequency Correction CHannel
FCS	Frame Check Sequence
FDD	Frequency Division Duplexing, technologie 4G utilisée en France notamment, avec 2 bandes de fréquences pour le send montant UL et le send descendant DL.
FDM	Frequency Division Multiplex
FDN	Fixed Dialling Number
FEC	Forward Error Correction
FER	Frame Erasure Ratio
FH	Frequency Hopping
FIB	Forward Indicator Bit
FIRIP	Fédération des Industriels des Réseaux d'Intiative Publique
FISU	Fill In Signal Units
FMC	Fixed Mobile Convergence, en IMS
FN	Frame Number
FQDN	Fully Qualified Domain Name, domain name du SLP émettant le SUPL INIT
FR	Full Rate
FSG	Foreign Subscriber Gateway
FSN	Forward Sequence Number
ftn	forwarded-to number
FTTH	Fiber To The Home, *the alternative is to use 4G to provide high bandwidth internet*
G	
GAA	Generic Authentication Architecture, *authentication with public-private key pairs and certificates.*

GBA	Generic Bootstrapping Architecture, authentication par secret partagé Ki entre SIM et HLR-HSS
GC1	Group Call 1. Proprietary Client-Application interface and protocol between a UE and an AS for the selection or management of a Multicast channel.
GCR	Group Call Register
GCS	Group Communication Service
GCSE	Group Communication Service Enablers
GCS	AS Group Communication Service Application Server. *Le MCPTT par exemple*
GDPR	General Data Protection Regulation, *english translation, see RGPD.*
GGSN	GPRS Gateway Support Node, in GPRS equipped network, provides the interface between an operator's own IP network and the external IP network (GRX mostly)
Gi	Interface to Internet from/to GGSN with normal IP packets, not encapsulated
GLR	Gateway Location Register. Acts as an HLR for the visitor to avoid UPDATE LOCATION being sent to the HPLMN. Used by a VPLMN to maintain a visitor in its own network.
GLONASS	GLObal'naya Navigatsionnaya Sputnikovaya Sistema (Global Navigation Satellite System)
GMLC	Gateway Mobile Location Center
GMLC	List List of the GT of the GMLCs authorized to perform a PSL
GMSC	Gateway Mobile-services Switching Center
GMSK	Gaussian Minimum Shift Keying (modulation)
GNSS	Global Navigation Satellite System
GPA	GSM PLMN Area
GPM	Global Permission Management
GPRS	General Packet Radio Service
GRX	The Intranet IP network used by mobile operators to exchange GPRS data. It is operated on a cooperative basis by the main international carriers.
GSA	GSM System Area
GSM	Global System for Mobile communications
GSM	MS GSM Mobile Station
GSM	PLMN GSM Public Land Mobile Network
GSM-R	GSM Railway adaptation (fast mobility)or
GT	Global Title (E164 numbering address)
GTP	Gprs Transfer Protocol, GTP-C for the signaling of contexts-sessions, GTP-U for the tunneling of packets, GTP for the accounting. GTPv1 for legacy 2G-3G and GTPv2 for 4G and recent 3G.
GTT	Global Title Translation
GUI	Graphic User Interface
GUTI	Globally Unique Temporary User Identity, includes the GUMMEI identifying the MME which allocated it, and the M-TMSI a temporary of the IMSI.

Gx	Interface (protocol DIAMETER Credit Control) between PCEF and PCRF
Gy	Real-time charging by the GGSN using a DIAMETER interface with the SDP
Gz	Off-line charging interface of the GGSN which transfers data tickets with the GTP' protocol

H

HADOPI	Haute Autorité pour la Diffusion des Oeuvres et la Protection des droits sur Internet, a french agency to fight the illegal downloading.
HAH	Haute Autorité du Hertz, a future French agency which will absorb HADOPI.
HALDE	Haute Autorité de Lutte contre les Discriminations pour l'Egalité, a French agency 2005-2011.
H-PCRF	Implementation of the PCRF in the HLPMN (Home access)
H-GMLC	Home Gateway Mobile Location Center IP address, used in Location services.
H-SLP	Home SLP, used in Location services
H223	Multiplexing protocol for visio, voice and control
H245	Control protocol in H223 for H324-M communications
H263	visio encoding standard
H264=MPEG4	visio encoding standard for H324-M
H265/HeVC	High Efficiency Video Codec. Follower of H264 with a more efficienc coding
H324-M	Standard for visio calls 3G using 64 kbits and ISUP
HANDO	HANDOver
HDLC	High level Data Link Control
HE	Home Equipment (le AuC du HLR-HSS en général)
HI1	Handover Interface Port 1, Commandes LEA-> PLMN
HI2	Handover Interface Port 2, Intercept Related Information PLMN → LEA
HI3	Handover Interface Port 3, Content of Communication PLMN → LEA
HLC	High Layer Compatibility
HLR	Home Location Register
HMAC	Hash Message Authentication
HMAC-MD5	A signature algorithm returning 16 octets for a string of any length. It uses a shared secret between the two entities exchanging messages
HOLD	Call hold supplementary service
HPLMN	Home PLMN
HPU	Hand Portable Unit
HR	Half Rate
HSN	Hopping Sequence Number
HSPDA	High Speed Downlink Packet Access (gives 250 Kbits useful data for visio)
HU	Home Units

I	
I-CSCF	Interrogating CSCF, en IMS le nœud qui interroge le HSS
i-SCSI	Internet Small Computer System Interface
I	Information frames (RLP)
IA	Incoming Access (closed user group SS)
IA5	International Alphabet 5
IAM	Initial Address Message
IAP	Internet Access Provider (Provides the access a modem or a permanent IP connection to the Internet, not necessarily a Portal or Content Provider)
IC	Interlock Code (CUG SS)
ICB	Incoming Calls Barred (within the CUG)
ICC	Integrated Circuit(s) Card
IC(pref)	Interlock Code of the preferential CUG
ICI	Interception Configuration Information
ICM	In-Call Modification
ID	IDentification/IDentity/IDentifier
IDN	Integrated Digital Network
IE	(signaling) Information Element
IEC	International Electrotechnical Commission
IEI	Information Element Identifier
IETF	Internet Engineering Task Force
I-ETS	Interim European Telecommunications Standard
IFC	Initial Filter Criteria, part of the User Data coded in XML returned in the Server-Assignment-Answer of the Cx protocol.
IGP	International Gateway Provider (SCCP access to the SS7 network)
IGW	International SCCP Gateway (synonym of IGP).
II-NNI	inter-IMS Network to Network Interface, *it is the SIP-T or SIP-I protocol*
IMEI	International Mobile station Equipment Identity
IMEISV	International Mobile station Equipment Identity with software version
IMPI	IP Multimedia Private Identity, *unique identity allocated to a user*
IMPU	IP Multimedia Public Identity, *there can be (a family) several IPMU for one IPMI*
IMS	IP Multimedia System
IMSI	International Mobile Subscriber Identity
IMSI	catcher, fake radio station with a strong signal used for the interception of mobile terminals.
IM-SSF	IP Multimedia Service Switching Function, SIP application interfacing SIP to Camel charging systems
IN	Interrogating Node
IN	Intelligent Network
IN	service Service such as pre-payment, number correction, APN correction performed by an SCP
INAP	Intelligent Network Application Part

INI	Internal Network Information. *Formats X12, X13 with the Mediation System for Legal Intercept.*
InitialDP	CAMEL service to start an IN Service
IOPTS	Isolated E-UTRAN Operator for Public Telecommunication Safety
IOT	Inter Operator Tarif
IoT	Internet of Things, *synonym of M2M*
IP	Internet Protocol
IP-CAN	IP Connectivity Access Network, *procedure to create, modify or delete bearer contexts.*
IP-PDP	Address IP allocated by the GGSN in the response to a Create PDP Context
IP-SM-GW	IP Short Message Gateway, *translates bith way MAP SMS messages 29.002 to SIP MESSAGE*
IP-TE-VPN	Address IP subsequently allocated by a RADIUS server to create a secure VPN tunnel
IR21	International Roaming 21document*(description of the detailed numbering plan as standardized by the GSM association)*
IRI	Intercept Related Information, *sent by the HI2 interface HI2 to the LEA's LEMF*
ISC	International Switching Center
ISD	Abbreviation for INSERT SUBSCRIBER DATA
ISD	Issuer Security Domain
ISD-R	Issuer Security Domain-Root
ISD-P	Issuer Security Domain-Profile
ISDN	Integrated Services Digital Network
ISO	International Organization for Standardization
ISP	Internet Service Provider (Content Provider for Internet services)
ISUP	ISDN User Part (of signaling system No.7)
ITC	Information Transfer Capability
ITU	International Telecommunication Union
IVR	Interactive Voice Response
IWF	InterWorking Function (used for Circuit Mode internet access, Modems V110-> packet converter)
IWF-SMS	Converter of SMS 4G (Diameter Sgd TS 29.338) - 3G (MAP TS 29.002)
IWMSC	InterWorking MSC
IWU	InterWorking Unit
J	
	.jav File of compiled Java source code into "byte-code"; to be executed by a Java VM platform independent interpreter.
K	
k	Windows size
K	Constraint length of the convolutional code
Kc	Ciphering key

Ki	Individual subscriber authentication key
KiC	Key for Ciphering (OTA SIM)
KiD	Key for RC/CC/DS signature (OTA SIM)
KiK	Key for protecting KiC and KiD
L	
L1	Layer 1
L2ML	Layer 2 Management Link
L2R	Layer 2 Relay
L2R	BOP L2R Bit Orientated Protocol
L2R	COP L2R Character Orientated Protocol
L3	Layer 3
LA	Location Area
LAC	Location Area Code
LAI	Location Area Identity
LAN	Local Area Network
LAPB	Link Access Protocol Balanced
LAPDm	Link Access Protocol on the Dm channel
LBO	Local Break-Out, allows local subscribers of an ARP (VPLMN) to use the data services.
LBS	Location-Based Services
LCN	Local Communication Network
LCS-AP	LCS Application Protocol
LCS	Location Services
LCSC	LCS Client
LCSS	LCS Server
LE	Local Exchange
LEA	Law Enforcement Agency
LEMF	Law Enforcement Monitoring Facility, *system at the LEA premises which monitors the IRIs*
LI	Lawful Interception
LI	Length Indicator -Line Identity
LIID	Lawful Interception Identifier
LIP	Location Information Platform
LIP	Location Information Protocol *Standard TETRA*
LIPA	Local IP Access, $>$ = Rel 10 LTE pour permettre l'interco directe entre équipements connectés à une femtocell ou au même eNodeB.
LIR	Location Information Request. En IMS l'équivalent d'un SEND ROUTING INFO FOR SM en MAP
	Lawful Interception
LLC	Low Layer Compatibility
Lm	Traffic channel with capacity lower than a Bm
LMSI	Local Mobile Station Identity, Normalisé ETSI pour TETRA
LMU	A Location Measurement Unit with only Air interface
LMU	B Location Measurement Unit integrated in a BTS.
LND	Last Number Dialled

LNP	Local Number Portability
LOCSIP	Location SIP. *Specified by OMA to include the localization in the SIP NOTIFY.*
LoRa	Long Range Radio
LPLMN	Local PLMN
LPA	Local Profile Update, *the SIM is updated by an application in the terminal*
LPP	LTE Positioning Protocol
LPPa	LTE Positioning Protocol Annex
LR	Location Registration
LSP	Location Service Provider, *MNO or third party.*
LSSU	Link Status Signal Units
LSTR	Listener SideTone Rating
LTE	Local Terminal Emulator
LTE	Long Term Evolution *(the new radio access standard of the 4G)*
LU	Local Units-Location Update
LV	Length and Value
M	
M	Mandatory
MA	Mobile Allocation
MAC	address Media Access Control address. *Unique identifier (manufacturer dependent) assigne to a physical network interface.*
MACN	Mobile Allocation Channel Number
MAF	Mobile Additional Function
MAH	Mobile Access Hunting supplementary service
MAI	Mobile Allocation Index
MAIO	Mobile Allocation Index Offset
MAP	Mobile Application Part, there is MAP GSM and MAP IS-41(CDMA)
MB2	Reference Point and protocol between the BM-SC and the Application Server. *MB2 CP is the Control Plane and MBMS UP is the user plane for the multicast user data*
MBMS	Multimedia Broadcast Multicast Services (TV broadcast on a single frequency per cell)
MBMS-GW	MBMS Gateway (can be integrated with the PGW)
MBSFN	Multicast Broadcast Single Frequency. *For the MBMS transmission, synchronization zone of the eNodeBs for the tightly synchronized transmission. The data are broacasted in all cells even if there are no concerned users.*
MC	Message Center *(in the IS 41 network, equivalent of a GSM SMSC)*
MCE	Multicast Coordination entity. *Unit which coordinates the multicast to several MBSFN.*
MCDATA	Mission Critical Data Control
MCVIDEO	Mission Critical Video services
Multi-Call	(simultaneous bearer services)
MCC	Mobile Country Code

MCCH	Multicast Control Channel, *eNodeB → UEs to broadcast the Multicast control on the entire cell coverage.*
MCE	Multicell/Multicast Coordination entity
MCH	MNP Clearing House
MCH	Multicast Channel
MCI	Malicious Call Identification supplementary service
MCPTT	Mission Critical Push-To-Talk
MTCH	Multicast Traffic Channel *eNodeN → UEs eNodeB → UEs to broadcast the Multicast data on the entire cell coverage.*
MD	Mediation Device
MD5	*A signature algorithm returning 16 octets for a string of any length including a shared secret at the end*
MDG	Mobile Data Gateway, *Interface Tetrapol with the geo-localization based on OMA MPL.*
MDL	(mobile) Management (entity) Data Link (layer)
MDM	Mobile Device Management, *Multi-terminal platform allowing to modify the system files of several handset manufacturers. MDM supplier examples: Airwatch, Blackberry BES, Mobile Iron, Pradeo (France).*
MDN	Mobile Destination Number (IS 41)
ME	Maintenance Entity
	Mobile Equipment
MEF	Maintenance Entity Function
MF	MultiFrame
	Mediation Function, *for Lawful Interception*
MFSMWR	MTU_FORWARD_SHORT_MSG_WWW_REQ
MFSMC	MTU_FORWARD_SHORT_MSG_CNF
MFSMHI	MTU_FORWARD_SHORT_MSG_HO_IND
MFSMWRSP	MTU_FORWARD_SHORT_MSG_WWW_RSP
MGIT	Message Gratuit d'Indication de Tarification, *ARCEP term for the mandatory tariff announcement free voice message before the charging of a service. It must be listened by the calling party for the approval procedure of an ISUP or SIP connection to a public operator.*
MGT	Mobile Global Title
MHS	Message Handling System
MIB	Master Information Block, *standard description of a system to be supervised by the SNMP protocol*
MIB	Master Information Block, *LTE, broadcasted for the paging procedure eNodeB → UE*
MIC	Mobile Interface Controller
Milenage	*A more secure equivalent 3G of the A5/A3 algorithm*
MIP	Mobile IP
MLC	Mobile Location Center
MLP	Mobile Location Protocol, XML Based independent of the technology or bearer service, OMA standard
MLU	Mobile Location Units

MM	Man Machine
	Mobility Management
MMD	Multi Media Domain
MME	Mobile Management Entity, equivalent 4G of an SGSN
MMI	Man Machine Interface
MMS	Multimedia Messaging Service
MM1	Protocols using IP standards(http) to exchange MMS between the cellphone and the MMSC
MM4	In the MMS architecture, protocol to send MMS from one MMSC to an other (interconnection), basically SMTP (e-mail).
MM5	In the MMS architecture, protocol to interrogate the HLRs
MM7	Protocols using IP standards so Content Provider sends MMS to an MMSC
MMT	Mobile Money Transfer
MNC	Mobile Network Code
MNO	Mobile Network Operators
MNP	Mobile Number Portability
MNP-SRF	Signaling Relay Function (a way to implement the MNP servive in an MNO)
MO	Mobile Originated
MOC	Mobile Originated Call
MO-LR	Mobile Originating Location Request
MOOC	Massive Open Online Course
MoU	Memorandum of Understanding
MPC	Mobile Positioning Center
MPEG	4 visio encoding standard for H324-M (alternative to H263 or H264)
MPH	(mobile) Management (entity) PHysical (layer) [primitive]
MPTY	MultiParTY (Multi ParTY) supplementary service
MRP	Mouth Reference Point
MS	Mobile Station
MSC	Mobile services Switching Center, Mobile Switching Center
Anchor	MSC Mobile Switching Center that is the first to assign a traffic channel to an MS
Serving	MSC MSC which currently has the MS obtaining service at one of its cell sites
Tandem	MSC Previous Serving MSC in the handoff chain
MSCM	Mobile Station Class Mark
MSCU	Mobile Station Control Unit
MSID	Mobile Station Identifier. Used for LBS
MSISDN	Mobile Station International ISDN Number
MSRN	Mobile Station Roaming Number
MSRP	Message Session Relay Protocol, *used with RCS to relay related instant messages and for file transfer*
MSU	Message Signal Units
MT	Mobile Terminated
MT (0,1,2)	-Mobile Termination

MT-LR	Mobile Terminating Location Request
MTC	Mobile Terminated Call
MTC	Machine Type Communication, *NB-IOT use*
MTC-IWF	MTC Interworking Function, *interfaces using T5 with, the MME on one side and the HSS with S6m/Diameter on the other side, also Tsp with the SCS.*
MTM	Mobile-To-Mobile (call)
MTN	Maintenance Regular Message
MTP	Message Transfer Part
MTP2	MTP Layer 2 (Link control level)
MTP3	MTP Layer 3 (Nework control sub-level (handles Point Codes)
MTRF	Mobile Terminating Roaming Forwarding, *In CFB procedure, improves the delay compared to MTR, by using the old MSC/VLR as relay.*
MTTR	Mobile Terminating Roaming Retry
MU	Mark Up
MULTICAST	"MULtiple broadCAST," broadcast such as what is used in Numerical TV (DVB-H). The content is broadcasted to any number of receivers like in classical TV thus much more economical than UNICAST currently used.
MUMS	Multi User Mobile Station
MVNO	Mobile Virtual Network Operator
MWD	Message Waiting Data (indication in an HLR)
M2M	Machine to Machine, *synonym of IoT*
M2PA	MTP2 Peer-to-Peer Adaptation Layer
M2UA	MTP2 User Adaptation Layer
M3UA	MTP3 User Adaptation Layer
N	
N/W	Network
NAMPS	Narrow AMPS
NAP	Net Assist Protocol, *Used in TETRA for the LBS services.*
NAPTR	Naming Authority Pointer: un enregistement particulier du DNS caractérisé par Application Service, Application Protocol et nom de domaine habituel.
NAS	Network Access System: *a Hot Spot, an SGSN or previously a modem rack for the data service using circuits.*
NAT	Network Address Translation (*a single external address for many local addresses*)
NB	Normal Burst
NBIN	A parameter in the hopping sequence
NBM	Network-Based Mobility
NB-IoT	Narrow-Band Internet of Things, *standard* for *3GPP IoT*
NCC	Network (PLMN) Country Code
NCELL	Neighboring (of current serving) Cell
NCH	Notification CHannel

NDC	Network Destination Code
NDUB	Network Determined User Busy
NE	Network Element
NEF	Network Element Function
NET	Norme Européenne de Télécommunications
NF	Network Function
NFC	Near Field Technology (*contactless communication of handset*)
NGN	Next GeNeration (IP-based equipment and networks)
NI	Network Indicator
NIC	Network Independent Clocking
NI-LR	Network Induced Location Request
NM	Network Management
NMC	Network Management Center
NMSI	National Mobile Station Identification number
Node B	The UMTS 3G equivalent of a BTS (GSM 2G)
NPI	Number Plan Indentifier
NPLR	Number Portability Location Register (implements the MNP-SRF function)
NPS	Network Planning System
NRF	Network Repository Function, 5G equivalent of an UDR?
NRH	Number Range Holder (the original network which has allocated the MSIDN)
NSAP	Network Service Access Point
NSS	Network Vendors
NT	Network Termination
	Non-Transparent
NTAAB	New Type Approval Advisory Board
NUA	Network User Access
NUI	Network User Identification
NUP	National User Part (SS7)
O	
O	Optional
OA	Outgoing Access (CUG SS)
	Origin Address
O&M	Operations & Maintenance
OACSU	Off Air Call SetUp
OCB	Outgoing Calls Barred within the CUG
OCS	Online Charging System
OCF	Online Charging Function, other name for OCS
OD	Optional for operators to implement for their aim
OFCS	Offline Charging System
OIV	Opérateur d'Importance Vitale (SNCF, EDF, etc. in France, for example)
OLR	Overall Loudness Rating
OMC	Operations & Maintenance Center

OML	Operations and Maintenance Link
OPC	Originating Point Code
OR	Optimal Routing
OS	Operating System
OSI	Open System Interconnection
OSI	RM OSI Reference Model
OSSS	Originating SMS Supplementary Service
OTA	Over the Air, *OTA SIM and OTA TDSA*
OTDOA	Observed Time Difference of Arrival
OTT	Over The Top, *service simple de VoIP sans mise en œuvre de bearers dédiés*
OVI	Operators Vitally Important, *such as public utilities, health, etc. which have an obligation of securing their telecommunications.*
P	
P-CSCF	Proxy Call Session Control Function, en IMS le relais dans le VPLMN avec interface Rx vers PCRF
PABX	Private Automatic Branch eXchange
PAD	Packet Assembly/Disassembly facility
PAP	Push Access Protocol. Uses an MIME format to send unsollicited XML data to a mobile device.
PCAP	Positioning Calculation Application Part
PCC	Policy Charging and Control
PCEF	Policy and Charging Enforcement Function. *Function in the PGW which is able to analyze (Deep Packet Inspection) the IP messages and enforces the rule that it has obtained from an associated PCRF*
PCell	"Primary Cell," *equivalent of the "Serving Cell"*
PCH	Paging Channel
PCI	Physical Cell ID, in 4G (*same rôle as PSC in 3G W-CDMA) scrambling code of a physical cell (cannot be the same for two neighbor cells which overlap)*
PCF	Policy Control Function, *5G equivalent of a PCRF*
PCM	Pulse Code Modulation
PCRF	Policy and Charging Control Function. *Equipment which registers the IP handling rules for specific customers and pass the rule to the PCEF. PCRF is like the SCP and the PCEF is like the SSP in a CAMEL architecture.*
PD	Protocol Discriminator
	Public Data
PDN	Public Data Networks
PER	Packet Encoding Rule, Variante du codage ASN1 utilisé pour les messages de la partie BSS
PGW	PDN Gateway (Packet Data Network Gateway), equivalent in 4G (LTE) of the GGSN. Uses GTP V2 (as latest 3G GGSN) or IPMI. Acts as an "anchor" between LTE and non-LTE (e.g., WiFi) access technology.

PH	Packet Handler
	PHysical (layer)
PHI	Packet Handler Interface
PI	Presentation Indicator
PICS	Protocol Implementation Conformance Statement
PIN	Personal Identification Number
PIXT	Protocol Implementation eXtra information for Testing
PLMN	Public Lands Mobile Network
PMD	Pseudonym Mediation Device Functionality
PMIP	Proxy Mobile IP, alternative non-roaming simpler protocol to LTE GTPv2 over the S5/S8 interface between an SGW and a PGW allowing a single tunnel.
PMR	Private Mobile Radio
PNE	Présentation des Normes Européennes
POI	Point Of Interconnection (with PSTN)
PoR	Proof of Receipt (OTA SIM)
POTAP	Push Over the Air Protocol. Based on WSP, a binary protocol of HTTP.
PP	Point-to-Point
PPE	Primative Procedure Entity
PPEP	Public Peering Exchange Point: *the main routers of the Internet and GRX networks. They can be common with different IP addresses which separate the two IP networks.*
PRACH	LTE Physical Random Access Channel (PRACH)
PPDR	Public Protection and Disaster Relief, *dedicated public safety telecom systems*
PPG	Push Proxy Gateway
PPR	Address Private Profile Register IP address, used with Location Services 3G, 4G
Pref	CUG Preferential CUG
PRN	Abbreviation for the MAP primitive Provide Roaming Number
ProSe	Proximity-Based Services 3GPP TS 23.303
PSL	MAP GSM service PROVIDE SUBSCRIBER LOCATION
PS	Domain Packet Service Domain (includes SGSN)
PSAP	Public Service Answering Points, such as emergency services 911(US) or 112(Europe)
PSC	Private Scrambling Code, in W-CDMA (3G) *identifies a cell.*
PSI	Public Service Identity, *identifies an AS*
PSPDN	Packet Switched Public Data Network
PSTN	Public Switched Telephone Network
PUCCH	Physical Uplink Control Chanel (used in LTE for the sending of the SRS by the UE)
PUCT	Price per Unit Currency Table
PW	Pass Word

Q

QAM	Quadrature Amplitude Modulation, *In LTE 256QAM (8 bits/symbol), 64QAM(6 bits/symbol); 16 QAM (4 bits/symbol) then QPSK (2 bits/symbol) when the SNR is lower.*
QoP	Quality of Position
QoS	Quality of Service
QPSK	Quadrature Phase Shift Keying, *2bits/symbol, used in LTE when the signal is lower than required for QAM.*

R

R	Value of Reduction of the MS transmitted RF power relative to the maximum allowed output power of the highest power class of MS (A)
RA	RAndom mode request information field
RAB	Random Access Burst
RAC	Routing Area Code *(for GPRS coverage)*
RACH	Random Access Channel
RADIUS	Protocol and Server *used for Authentication-Authorization and Accounting purposes*
RAI	Routing Area Information *(for GPRS coverage)*
RAN	Radio Access Network
RANAP	Radio Access Network Application Part *(equivalent 3G of BSSAP 2G) on the Iu interface CN-RNC*
RAND	*RANDom number (used for authentication)*
RBER	Residual Bit Error Ratio
RCP	Remote Control Point *(the SCCP function of an MSC (Alcatel term)*
RCS	Rich Communication Suite. *Uses extensions of the SIP protocol for chat and file transfers*
RDI	Restricted Digital Information
REC	RECommendation
REJ	REJect(ion)
REL	RELease
REQ	REQuest
RF	Radio Frequency
RFC	Radio Frequency Channel
RFCH	Radio Frequency CHannel
RFN	Reduced TDMA Frame Number
RFU	Reserved for Future Use
RGPD	Règlement Général sur la Protection des Données, *EU law applicable the 25 may 2018 protecting private data in particular forbidding the transfer of European resident data outside the EU*
RH	Roaming Hub
RLP	Radio Link Protocol
RLP	Roaming Location Protocol, between a V-SLP and an H-SLP
RLR	Receiver Loudness Rating
RLS	LOCSIP Resource Location Server
RMS	Root Mean Square (value)

RN	Routing Number (the network "prefix" to transform a given number subject to MNP)
RNC	Radio Network Controller UMTS 3G (equivalent to BSC (GSM 2G)
RNTABLE	Table of 128 integers in the hopping sequence
ROSE	Remote Operation Service Element
RP	Reply Path
RPOA	Recognized Private Operating Agency
RPLMN	Registered PLMN, *the last VPMLN which a mobile registered on.*
RR	Radio Resource
RRC	Radio Resource Control. *Supports the exchanges between the handset and SMLC or SLP.*
RRLP	Radio Resource LCS Protocol. *For 2G and 3G same role as LPP for 4G.*
RSA	Rivest, Shamir, Adleman, *public key asymetrical ciphering system. The patent is in the public doamin since 2000 and is used in particular to exchange symetrical ciphering keys.*
RSE	Radio System Entity
RSL	Radio Signaling Link
RSRP	Reference Signal Received Power. *One of the measurements provided by the eNodeB to the SMLC with the ECID localization method. See TS 36.255*
RSRQ	Reference Signal Received Quality. *One of the measurements provided by the eNodeB to the SMLC with the ECID localization method, the Signal/Noise Ratio SNR. See TS 36.255*
RSZI	Regional Subscription Zone Identity
RTD	Real Time Difference *(used un hyperbolic location methods)*
RTE	Remote Terminal Emulator
RTCP	Real Time Control Protocol
RTP	Real Time Protocol *(used to carry 3G H323 visio sessions)*
RTSP	Real Time Streaming Protocol *(distributes a multimedia stream (e.g.TV)* on an IP connection.
RTT	Round Trip Time
RTTH	Radio To The Home, *Use of 4G local loop to provide the internet high bandwidth access*
RXLEV	Received signal level
RXQUAL	Received Signal Quality
S	
S-CSCF	Serving CSCF, en IMS
S/W	SoftWare
SABM	Set Asynchronous Balanced Mode
SACCH	Slow Associated Control Channel
SACCH/C4	Slow Associated Control CHannel/SDCCH/4
SACCH/C8	Slow Associated Control CHannel/SDCCH/8
SACCH/T	Slow Associated Control CHannel/Traffic channel
SACCH/TF	Slow Associated Control CHannel/Traffic channel Full rate

SACCH/TH	Slow Associated Control CHannel/Traffic channel Half rate
SACH	Service Announcement Channel
SAE	System Architecture Evolution, *Architecture of the LTE network*
SAI	Support of Area Identity
SAI	Service Area Identity, in the MBMS service the identity of an MBSFN area
SAN	Storage Area Network
SAP	Service Access Point
SAPI	Service Access Point Indicator
SAS	Stand Alone SMLC (*in 3G connected by the Iupc interface to an SRNC*)
SB	Synchronization Burst
SC	Service Center (used for SMS)
	Service Code
SC-PTM	Single Cell Point To Multipoint transmission. Alternative to MBMS, the Multicast data are broadcasted only in cells where there are concerned users
SCE	Service Creation Environment (scripting for IN or IVR or USSD)
SCEF	Service Capability Exposure Function, *the PGW equivalent for IoT services with the S6t interface with the HSS.*
SCP	Service Control Point. In an IN system, this is the SS7 part which implements the various service logics (*pre-paid, number correction, anti-tromboning, etc.*)
SCCP	Signaling Connection Control Part
SCEP	Server Key Enrolment Protocol, CISCO protocol for registering PKI certificates in particular iPhones.
SCF	Service Control Function
SCH	Synchronization Channel
SCLC	SCCP Connection Less Control
SCMG	SCCP Management
SCN	Subchannel Number
SCOC	SCCP Connection Oriented Control
SCF	Service Control Function
SCP	Service Control Point
SCRC	SCCP Routing Control
SCS	Service Capability Server, *a server used in the NB-IOT architecture*
SCTP	Stream Control Transmission Protocol (*TCP with multi-homing*)
SDCCH	Stand-alone Dedicated Control CHannel
SDL	Specification Description Language
SDP	Service Data Point. *In an IN system this is the database with the subscriber records. It is interrogated by the SCP using the DIAMETER protocol*
SDP	Session Description Protocol, used with RCS
SDS	Short Data Service, *TETRA equivalent of GSM SMS service*
SDT	SDL Development Tool
SDU	Service Data Unit

SE	Support Entity
SGSN	Support GPRS Service Node. In GSM 2.5 G with GPRS, it has both circuit and IP interfaces, and provides the GPRS service to a visiting cellphone. It can deliver SMS-MT
SEF	Support Entity Function
SET	SUPL Enabled Terminal
	AreaSF Status Field
SFH	Slow Frequency Hopping
SFN	Single-Frequency Network, By MBMS A synchronization zone of the eNodeBs
SGi	Interface to Internet from/to a LTE PGW with normal IP packets, not encapsulated (*equivalent to Gi for a 3G GGSN*).
SGi-mb	User data interface between BM-SC and MBMS GW for the multicast service
SGP	SIGTRAN Gateway Point (IP< -> TDM conversion)
SGP	Voice management system, Gateway with PSTN for Tetrapol
SGSN	Serving GPRS Support Node, *node which handles the PS domain in 2G or 3G*
SI	Screening Indicator
Si	*IMS Interface with the IM-SSF to transport Camel subscription information*
SIB	System Information Block, LTE, dans la procédure de paging eNodeB-> UE
	Service Interworking
	Supplementary Information
SID	SIlence Descriptor
SIGTRAN	Signal Transport (Working Group which works on SS7/IP)
SIF	Signaling Information Field
SIM	Subscriber Identity Module
SIO	Service Information Octet
SIP	Session Initiated Protocol (the VoIP protocol)
SIPTO	Selected Internet IP Traffic Offload LTE > = Rel 10
SLC	Signaling Link Code
SLIA	Standard Location Immediate Answer of the MLP protocol
SLIR	Standard Location Immediate Request of the MLP Protocol
SLF	Server Locator Function, *allows an HSS to find another one*
SLP	SUPL Location Platform
SLPP	Subscriber LCS Privacy Profile
SLR	Send Loudness Rating
SLS	Signaling Link Selection
SLTA	Signaling Link Test Message Acknowledgment
SLTM	Signaling Link Test Message (*polling between adjacent Point Codes*)
Sm	GTPv2 interface between the MME and the eMBMS-GW, *a MME must have it to be multicast service compatible*
SM	Short Message
SM	Subscription Manager

SM-SR	Subscription Manager-Secured Routing
SM-DP	Subscription Manager-Data Preparation
SME	Short Message Entity
SMF	Session Management Function, *a 5G equipment in TS 23.501, which regroups the handling of the Control Plane of SGW and PGW*
SMIL	Synchronized Multimedia Integration Language (*to code animated sequences for MMS*)
SMG	Special Mobile Group
SMLC	Serving Mobile Location Center
SMPP	Short Message Point to Point, OMA standardized originally developed by Logica.
SMS	Short Message Service
SMSC	Short Message Service Center
SMSCB	Short Message Service Cell Broadcast
SMSDPTP	SMS Delivery Point to Point
SMSDBCKW	SMS Delivery Backward
SMSDFWD	SMS Delivery Forward
SMS-SC	Short Message Service-Service Center
SMS/PP	Short Message Service/Point-to-Point
Smt	Short message terminal
SM-AL	Short Message Application Layer
SM-TL	Short Message Transfer Layer
SM-RL	Short Message Relay Layer
SM-RP	Short Message Relay Protocol
SMN	Shared MBMS Network
SN	Subscriber Number
SNM	Signaling Network Management
SNMP	Small Network Management Protocol (*uses an MIB description of the equipment)*
SNR	Serial Number
SNR	Signal/Noise Ratio, same as RSRQ
SNRC	Serving RNC (in a 3G network)
SOA	Suppress Outgoing Access (CUG SS)
SoLSA	Support of Localized Service Areas
SON	Self Organizing Networks
SOR	Steering Of Roaming
SP	Service Provider
	Signaling Point
	SPare
SPC	Signaling Point Code
SPC	Suppress Preferential CUG
SPC	Semi Persistent Scheduling
SRES	Signed RESponse (authentication)
SRI	Abbreviation for the MAP primitive Send_Routing_Info
	SRI_FOR_SM Abbreviation for the MAP primitive Send Routing Info for Short Message

SRF	Service Resource Function
SRNS	Serving Radio Network Subsystem (GPRS)
SRS	Sounding Reference Signals. The messages sent by a UE and received by the LMU in the U-TDOA positioning method.
SS	Supplementary Service
	System Simulator
SSC	Supplementary Service Control string
SSF	Subservice Field
	Service Switching function (the IN part of an MSC, it may have its own GT)
SSM	Source Specific Multicast
SSN	Sub-System Number
SST	Sub-System Test (polling between SCCP sub-systems)
SSTA	Sub-System Test Acknowledgment
SS7	Signaling System No. 7
SSP	Service Switching Point
STA	Station, l'UE, le MS ou le terminal WiFi
STMR	SideTone Masking Rating
STP	Signaling Transfer Point
Stut	Expression belge "il y a un stut" équivalent à "il y a un loup"
SU	Signal Unit
SUA	SCCP User Adaptation Layer
	Supplicant The low level software layer in a terminal which establishes the IP link through the Access Point.
SUPL	Secure User-Plane Location protocol
SVN	Software Version Number
SVI	Serveur Vocal Interactif
SWm	Interface (Diameter) between the AAA server and the ePDG
SWP	Single Wire Protocol
SWx	Interface between HSS and AAA server with a registration-cancellation protocol on a WLAN
T	
T	Timer
	Transparent
	Type only
TA	Terminal Adaptor
TAC	Tracking Are Code
TAI	Tracking Area Identity, *4G (16 bits), Identify the Area where the UE is located (used for Handover and CFB)*
TAP	Transferred Account Procedure, *currenly V3.12 is used for the transfer of charging tickets for roaming.*
TC	Transaction Capabilities
TCAP	Transaction Capability Application Part
TCH	Traffic CHannel
TCH/F	A full rate TCH
TCH/F2,4	A full rate data TCH (2,4 kbit/s)

TCH/F4,8	A full rate date TCH (4,8 kbit/s)
TCH/F9,6	A full rate data TCH (9,6 kbit/s)
TCH/FS	A full rate Speech TCH
TCH/H	A half rate TCH
TCH/H2,4	A half rate data TCH (2,4 kbit/s)
TCH/H4,8	A half rate data TCH (4,8 kbit/s)
TCH/HS	A half rate Speech TCH
TCI	Transceiver Control Interface
TC-TR	Technical Committee Technical Report
TCS	TETRA Connectivity Server, *Airbus proprietary protocol for sending/receiving messages with the SDS server*
TDD	Time Division Duplexing, 4G technology with a single frequency band and dynamic allocation of time slices for the UL and DL senses
TDF	Traffic Detection Function, dans un PCEF
TDMA	Time Division Multiple Access
TDSA	Terminal Data Service Access, acronym for OTA TDSA (equivalent of OTA GPRS to load APNs, etc.)
TE	Terminal Equipment
TEID	Terminal endpoint identifier, Identity of a tunnel between SGSN-GGSN or MS-PGW
TETRA	Terrestrial Trunked Radio, PMR Technology normalized by ETSI
Tetrapol	PMR Technology used for private security forces networks with services like Push to Talk and group calls, not normalized by ETSI
TFA	TransFer Allowed
TFP	TransFer Prohibited
TFT	Traffic Flow Template, *Filter system on IPs and ports allowing an EU or PGW to route traffic to different "bearers"*
THD	Très haut Débit
TI or TID	Transaction Identifier (*in the TCAP protocol*)
	TLV Type, Length and Value
TMGI	Temporary Mobile Group Identity, *allocated for the MBMS service*
TMN	Telecommunications Management Network
TMSI	Temporary Mobile Subscriber Identity
TN	Timeslot Number
TOA	Time of Arrival
TOE	Target of Evaluation
TON	Type Of Number
TOS	Type of Service *A useful 6-bit field in the IP header giving the required transport quality. Replaced by DSCP which provided 2 additional bits.*
TP	Transfer Protocol (in the MAP protocol)
TPF	Traffic Plane Function
TRX	Transceiver
TS	Time Slot
Ts	Basic Timing unit 4G, which is $1/(15000 \times 2048) = 32.55$ ns.
	Technical Specification
	TeleService

TSC	Training Sequence Code
TSDI	Transceiver Speech & Data Interface
TTCN	Tree and Tabular Combined Notation, language de description normalisé ETSI de séquences de test d'équipements.
TUA	TCAP User Adaptation Layer
TUP	Telephone User Part (SS7)
TV	Type and Value
TXPWR	Transmit PoWeR; Tx power level in the MS_TXPWR_REQUEST and MS_TXPWR_CONF parameters
TWAN	Trusted WLAN Access Network
TWLAN	Trusted WLAN, secured WLAN used for non-LTE access to data services from a PGW through the S2a interface
U	
UA	User Adaptation
UDI	Unrestricted Digital Information
UDT	Unit Data Message (*of SCCP*)
UDUB	User Determined User Busy
UDM	Unified Data Management, *5G equivalent of a 4G HSS (the signaling part without the AuC)*
UDR	User Data Repository, an intermediate database between the CRM Provisioning and the HLR-HSS and PCRF. An element of the 5G core network architecture.
UE	User Equipment, equivalent to MS in 4G
UI	Unnumbered Information (*Frame*)
UIC	Union Internationale des Chemins de Fer
UICC	Cartes SIM dernière génération
UL	Abbreviation for UPDATE LOCATION
UL	Uplink
UL	RTOA Uplink Relative Time Of Arrival, used in the U-TDOA positioning method
UMA	Unlicensed Mobile Access
UNICAST	"UNIque broadCASTing": TV broadcast method where each reception has a dedicated channel (*can be used for "on demand"*) and the content is "streamed".
Unicast	Shared Key between Supplicant SW in the terminal and the Access Point to secure the Radio channel.
UPCMI	Uniform PCM Interface (13-bit)
UPF	User Plane Function, *5G function is TS 23.501 which regroups the handling of the User Plane for SGW and PGW.*
UPD	Up to date
USSD	Unstructured Supplementary Service Data
UUS	User-to-User Signaling supplementary service
UTRAN	UMTS Terrestrial Radio Access Network (3G) equivalent to BSS (2G)
U-TDOA	Uplink-Time Difference of Arrival

V	
V	PCRF Implementation of the PCRF in the VLPMN
VAD	Voice Activity Detection
VAP	Videotex Access Point
VBS	Voice Broadcast Service
VGCS	Voice Group Call Service
VLR	Visitor Location Register
VMS	Voice Mail System
VMSC	Visited MSC
VPLMN	Visited PLMN
VPN	Virtual Private Network
VoIP	Voice Over IP protocol
VoWifi	VoIP using a VPN IPsec to attach to an ePDG *(see chapter 8)*
VSC	Videotex Service Center
VCSG	Visited MSC or SGSN for a CSG
V(SD)	Send state variable
VTX	host The components dedicated to Videotex service

W	
WAN	Wide Area Network
WAP	Wireless Application Protocol
WBXML	Wireless "Binary" XML
WML	Wireless Markup Language
WLL	Wireless Local Loop
WLNP	Wireless Local Number Portability
Wm	Interface in LTE between the ePDG and the AAA server
WS	Work Station
WSS	Web Self-Service
WPA	Wrong Password Attempts *(counter)*

X	
X1	LI interface 1 to the PLMN equipements, specific HI1 interface for MSC, HLR, etc.
X2	LI interface 2 from the PLMN equipements, specific HI2 interface from MSC, HLR, etc.
X3	LI interface 3 from the PLMN equipements, specific HI3 interface from MSC, HLR, etc.
XID	eXchange Identifier
XML	Extensible Markup Language
XRES	Expected Response *(from the USIM card when computing A3)*

Z	
ZBLR	Zero Ball Loss Rake, *rake system to recycle balls used to clean steam condenser water pipes, nothing to do with Telecoms, just to lighten the end of the acronym list and fill letter Z*
ZC	Zone Code

Index

About the Author

Arnaud Henry-Labordère graduated from École Centrale de Paris (1966) and had the Ph.D. Mathematics, USA (1968).

Academic:

Was the professor of Operations Research at Ecole Nationale des Ponts et Chaussées from 1981 to 1998, then honorary professor after being associate professor from 1973 to 1980. He was also professor at Ecole Nationale des Mines de Paris, visiting professor at the Université de Versailles PRISM-CNRS, and previously assistant professor at Ecole Centrale de Paris and Université Paris Dauphine.

Industry:

Started as a mathematician at IBM USA in 1967 (Ph.D. fellowship). Then senior engineer at SEMA and SESA being project manager for the ground segment software of METEOSAT the first european ESA geostationnary satellite. 2 years at SITA as deputy technical directror designing a novel world wide telecom network. Has founded 3 companies: Ferma (1983) manufacturer of voice processing and large voice mail systems, devopping text-to-speech and voice recognition systems, then Nilcom (1998) the first company offering a world wide SMS network. He created Halys in 2003 as Chairman and Chief scientist, a mobile core network and virtual roaming systems manufacturer including the cyber security applications.

Author of 12 books (6 in Applied Maths, 5 in Telecommunications, 1 naval history), about 90 patents, 60 scientific articles and conferences.